Selected Titles in This Series

694 **Alberto Bressan, Graziano Crasta, and Benedetto Piccoli,** Well-posedness of the Cauchy problem for $n \times n$ systems of conservation laws, 2000

693 **Doug Pickrell,** Invariant measures for unitary groups associated to Kac-Moody Lie algebras, 2000

692 **Mara D. Neusel,** Inverse invariant theory and Steenrod operations, 2000

691 **Bruce Hughes and Stratos Prassidis,** Control and relaxation over the circle, 2000

690 **Robert Rumely, Chi Fong Lau, and Robert Varley,** Existence of the sectional capacity, 2000

689 **M. A. Dickmann and F. Miraglia,** Special groups: Boolean-theoretic methods in the theory of quadratic forms, 2000

688 **Piotr Hajłasz and Pekka Koskela,** Sobolev met Poincaré, 2000

687 **Guy David and Stephen Semmes,** Uniform rectifiability and quasiminimizing sets of arbitrary codimension, 2000

686 **L. Gaunce Lewis, Jr.,** Splitting theorems for certain equivariant spectra, 2000

685 **Jean-Luc Joly, Guy Metivier, and Jeffrey Rauch,** Caustics for dissipative semilinear oscillations, 2000

684 **Harvey I. Blau, Bangteng Xu, Z. Arad, E. Fisman, V. Miloslavsky, and M. Muzychuk,** Homogeneous integral table algebras of degree three: A trilogy, 2000

683 **Serge Bouc,** Non-additive exact functors and tensor induction for Mackey functors, 2000

682 **Martin Majewski,** ational homotopical models and uniqueness, 2000

681 **David P. Blecher, Paul S. Muhly, and Vern I. Paulsen,** Categories of operator modules (Morita equivalence and projective modules, 2000

680 **Joachim Zacharias,** Continuous tensor products and Arveson's spectral C^*-algebras, 2000

679 **Y. A. Abramovich and A. K. Kitover,** Inverses of disjointness preserving operators, 2000

678 **Wilhelm Stannat,** The theory of generalized Dirichlet forms and its applications in analysis and stochastics, 1999

677 **Volodymyr V. Lyubashenko,** Squared Hopf algebras, 1999

676 **S. Strelitz,** Asymptotics for solutions of linear differential equations having turning points with applications, 1999

675 **Michael B. Marcus and Jay Rosen,** Renormalized self-intersection local times and Wick power chaos processes, 1999

674 **R. Lawther and D. M. Testerman,** A_1 subgroups of exceptional algebraic groups, 1999

673 **John Lott,** Diffeomorphisms and noncommutative analytic torsion, 1999

672 **Yael Karshon,** Periodic Hamiltonian flows on four dimensional manifolds, 1999

671 **Andrzej Rosłanowski and Saharon Shelah,** Norms on possibilities I: Forcing with trees and creatures, 1999

670 **Steve Jackson,** A computation of δ^1_5, 1999

669 **Seán Keel and James M^c^Kernan,** Rational curves on quasi-projective surfaces, 1999

668 **E. N. Dancer and P. Poláčik,** Realization of vector fields and dynamics of spatially homogeneous parabolic equations, 1999

667 **Ethan Akin,** Simplicial dynamical systems, 1999

666 **Mark Hovey and Neil P. Strickland,** Morava K-theories and localisation, 1999

665 **George Lawrence Ashline,** The defect relation of meromorphic maps on parabolic manifolds, 1999

664 **Xia Chen,** Limit theorems for functionals of ergodic Markov chains with general state space, 1999

(*Continued in the back of this publication*)

Well-Posedness of the Cauchy Problem for $n \times n$ Systems of Conservation Laws

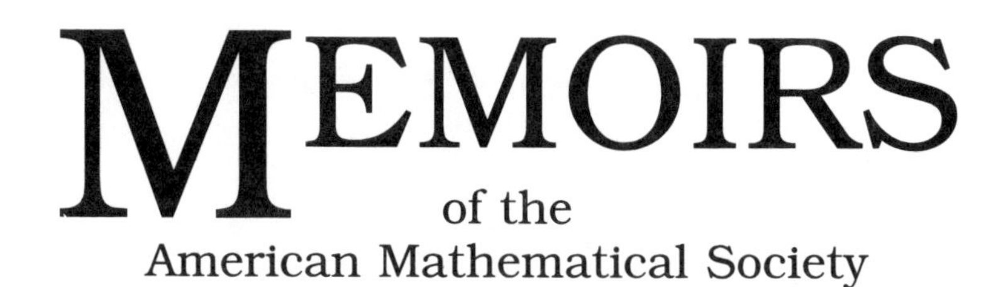

Number 694

Well-Posedness of the Cauchy Problem for $n \times n$ Systems of Conservation Laws

Alberto Bressan
Graziano Crasta
Benedetto Piccoli

July 2000 • Volume 146 • Number 694 (third of 5 numbers) • ISSN 0065-9266

American Mathematical Society
Providence, Rhode Island

2000 *Mathematics Subject Classification.*
Primary 35L65.

Library of Congress Cataloging-in-Publication Data

Bressan, Alberto, 1956–
Well-posedness of the Cauchy problem for $n \times n$ systems of conservation laws / Alberto Bressan, Graziano Crasta, Benedetto Piccoli.
p. cm. — (Memoirs of the American Mathematical Society, ISSN 0065-9266 ; no. 694)
Includes bibliographical references.
ISBN 0-8218-2066-4
1. Cauchy problem 2. Conservation laws (Mathematics) I. Crasta, Graziano. II. Piccoli, Benedetto, 1968– III. Title. IV. Series.
QA3.A57 no. 694
[QA377]
510 s—dc21
[515′.353] 00-036257

Memoirs of the American Mathematical Society

This journal is devoted entirely to research in pure and applied mathematics.

Subscription information. The 2000 subscription begins with volume 143 and consists of six mailings, each containing one or more numbers. Subscription prices for 2000 are $466 list, $419 institutional member. A late charge of 10% of the subscription price will be imposed on orders received from nonmembers after January 1 of the subscription year. Subscribers outside the United States and India must pay a postage surcharge of $30; subscribers in India must pay a postage surcharge of $43. Expedited delivery to destinations in North America $35; elsewhere $130. Each number may be ordered separately; *please specify number* when ordering an individual number. For prices and titles of recently released numbers, see the New Publications sections of the *Notices of the American Mathematical Society.*

Back number information. For back issues see the *AMS Catalog of Publications.*

Subscriptions and orders should be addressed to the American Mathematical Society, P. O. Box 5904, Boston, MA 02206-5904. *All orders must be accompanied by payment.* Other correspondence should be addressed to Box 6248, Providence, RI 02940-6248.

Memoirs of the American Mathematical Society is published bimonthly (each volume consisting usually of more than one number) by the American Mathematical Society at 201 Charles Street, Providence, RI 02904-2294. Periodicals postage paid at Providence, RI. Postmaster: Send address changes to Memoirs, American Mathematical Society, P. O. Box 6248, Providence, RI 02940-6248.

This publication is indexed in *Science Citation Index*®, *SciSearch*®, *Research Alert*®, *CompuMath Citation Index*®, *Current Contents*®*/Physical, Chemical & Earth Sciences.*
Printed in the United States of America.

♾ The paper used in this book is acid-free and falls within the guidelines established to ensure permanence and durability.
Visit the AMS home page at URL: `http://www.ams.org/`

10 9 8 7 6 5 4 3 2 1 05 04 03 02 01 00

Contents

ABSTRACT. This paper is concerned with the initial value problem for a strictly hyperbolic $n \times n$ system of conservation laws in one space dimension:

$$(*) \qquad u_t + \left[F(u)\right]_x = 0, \qquad u(0,x) = \bar{u}(x).$$

Each characteristic field is assumed to be either linearly degenerate or genuinely nonlinear. We prove that there exist a domain $\mathcal{D} \subset \mathbf{L}^1$, containing all functions with sufficiently small total variation, and a uniformly Lipschitz continuous semigroup $S\colon \mathcal{D} \times [0,\infty[\, \mapsto \mathcal{D}$ with the following properties. Every trajectory $t \mapsto u(t,\cdot) = S_t\bar{u}$ of the semigroup is a weak, entropy-admissible solution of $(*)$. Viceversa, if a piecewise Lipschitz, entropic solution $u = u(t,x)$ of $(*)$ exists for $t \in [0,T]$, then it coincides with the semigroup trajectory, i.e. $u(t,\cdot) = S_t\bar{u}$. For a given domain $\mathcal{D}$, the semigroup S with the above properties is unique.

These results yield the uniqueness, continuous dependence and global stability of weak, entropy-admissible solutions of the Cauchy problem $(*)$, for general $n \times n$ systems of conservation laws, with small initial data.

Received by the editor February 25, 1997; and in revised form March 9, 1998.

CHAPTER 1

Introduction

Consider the Cauchy problem for a strictly hyperbolic $n \times n$ system of conservation laws in one space dimension:

$$u_t + \big[F(u)\big]_x = 0, \tag{1.1}$$

$$u(0,x) = \bar{u}(x). \tag{1.2}$$

For initial data $\bar{u} \in \mathbf{L}^1$ with small total variation, a well known theorem of Glimm [**G**] provides the global existence of weak solutions. Aim of the present paper is to show that these solutions are unique and depend continuously on the initial conditions, with a Lipschitz constant in $\mathbf{L}^1$ which is uniform w.r.t. time. More precisely, the following holds.

THEOREM 1.1. *Let $\Omega \subseteq \mathbb{R}^n$ be an open set containing the origin, and let $F\colon \Omega \mapsto \mathbb{R}^n$ be a smooth map. Assume that the system (1.1) is strictly hyperbolic and that each characteristic field is either linearly degenerate or genuinely nonlinear. Then there exist a closed domain $\mathcal{D} \subset \mathbf{L}^1(\mathbb{R};\ \mathbb{R}^n)$, constants η_0, L, and a continuous semigroup $S\colon \mathcal{D} \times [0,\infty[\mapsto \mathcal{D}$ with the properties:*

(i) *Every function $\bar{u} \in \mathbf{L}^1$ with Tot. Var.$(\bar{u}) \leq \eta_0$ lies in $\mathcal{D}$.*
(ii) *For all $\bar{u}, \bar{v} \in \mathcal{D}$, $t, s \geq 0$ one has $\left\|S_t\bar{u} - S_s\bar{v}\right\|_{\mathbf{L}^1} \leq L\big(|t-s| + \left\|\bar{u} - \bar{v}\right\|_{\mathbf{L}^1}\big)$.*
(iii) *If $\bar{u} \in \mathcal{D}$ is piecewise constant, then for $t > 0$ sufficiently small the function $u(t,\cdot) = S_t\bar{u}$ coincides with the solution of (1.1)-(1.2) obtained by piecing together the standard self-similar solutions of the corresponding Riemann problems.*

The positively invariant domain $\mathcal{D}$ will have the form

$$\mathcal{D} = cl\left\{u \in \mathbf{L}^1(\mathbb{R};\mathbb{R}^n);\ u \text{ is piecewise constant},\ V(u) + C\cdot Q(u) < \delta_0\right\}, \tag{1.3}$$

for some constants $C, \delta_0 > 0$. Here $V(u)$ and $Q(u)$ denote the total strength of waves and the wave interaction potential of u, while cl denotes closure.

Following [**B5**], we say that a map S with the properties (i)–(iii) is a *Standard Riemann Semigroup* (SRS). The existence of such a semigroup was proved in [**B1, B3**] for special classes of $n\times n$ systems with coinciding shock and rarefaction curves, and in [**B-C1**] for general 2×2 systems. The present theorem, dealing with general $n \times n$ systems, contains all previous results in this direction.

Observe that the statement of Theorem 1.1 does not explicitly say that the trajectories of the semigroup are actually weak solutions of (1.1). This fact, however, can be deduced as a consequence of (i)–(iii), together with a number of additional properties which are collected below.

THEOREM 1.2. *For a given domain $\mathcal{D}$ of the form (1.3), there can be at most one continuous semigroup $S\colon \mathcal{D}\times[0,\infty[\,\mapsto \mathcal{D}$ satisfying the conditions (i)–(iii) listed in Theorem 1.1. If a SRS does exist, then the following properties also hold:*

(iv) *Each trajectory $t\mapsto u(t,\cdot) = S_t\bar{u}$ is a weak, entropy-admissible solution of the corresponding Cauchy problem (1.1)-(1.2).*

(v) *Let $(u_\nu)_{\nu\geq 1}$ be a sequence of approximate solutions of (1.1)-(1.2) generated by a wave-front tracking algorithm, or by the Glimm scheme with uniformly distributed sampling. Then, as $\nu\to\infty$ we have $\mathbf{L}^1$-$\lim u_\nu(t,\cdot) = S_t\bar{u}$ for every $t\geq 0$.*

(vi) *Let $u = u(t,x)$ be a piecewise Lipschitz, entropic solution of (1.1)-(1.2) defined on some strip $[0,T]\times\mathbb{R}$. Then $u(t,\cdot) = S_t\bar{u}$ for all $t\in[0,T]$.*

A proof of Theorem 1.2 is contained in [**B5**]. In turn, from the existence of the semigroup, one can prove the uniqueness of the entropy-weak solution of a given Cauchy problem, under a mild assumption on the growth of the total variation [**B-LF1**] or on the decay of positive waves [**B-G**]. Another application which is worth mentioning is the error estimate [**B-M2**], concerning the $\mathbf{L}^1$ distance between the exact solution of a Cauchy problem and an approximate solution generated by the Glimm scheme with uniformly distributed sampling [**L**].

There are three types of estimates which play a fundamental role in our analysis. These are: the estimates of Glimm on the total strength of waves [**G, Sm**], the local integral estimates used in the definition of Viscosity Solution [**B5**], and the decay estimates for positive waves of genuinely nonlinear families [**B-C3**].

A precise statement of these a-priori bounds requires some notation. Call $A(u) = DF(u)$ the Jacobian matrix of F at u. Smooth solutions of (1.1) thus satisfy the equivalent quasilinear system

$$u_t + A(u)u_x = 0. \tag{1.4}$$

Let $\lambda_1(u) < \cdots < \lambda_n(u)$ be the eigenvalues of $A(u)$ and choose right and left eigenvectors $r_i(u)$, $l_i(u)$, $i = 1,\ldots,n$, normalized so that

$$\nabla\lambda_i\cdot r_i(u) \doteq \lim_{h\to 0}\frac{\lambda_i\big(u+hr_i(u)\big)-\lambda_i(u)}{h} \geq 0, \tag{1.5}$$

$$|r_i|\equiv 1, \qquad \langle l_i,\, r_j\rangle = \begin{cases}1 & \text{if } i=j,\\ 0 & \text{if } i\neq j.\end{cases} \tag{1.6}$$

Following [**B-C2, Sch**], we now extend the definition of the Glimm functional to a general BV function. Let $u\colon\mathbb{R}\mapsto\mathbb{R}^n$ have bounded variation. Then $\mu\doteq D_x u$ is a vector measure, which can be decomposed into a continuous and an atomic part: $\mu = \mu^c+\mu^a$. For $i=1,\ldots,n$, define the signed measure $\mu_i = \mu_i^c+\mu_i^a$ as follows. The continuous part of μ_i is the Radon measure such that

$$\int\phi\, d\mu_i^c = \int l_i(u)\cdot\phi\, d\mu^c$$

for every scalar continuous function ϕ with compact support. The atomic part of μ_i is the measure concentrated on the countable set $\{x_\alpha;\ \alpha = 1,2,\ldots\}$ where u has a jump, such that $\mu_i^a\big(\{x_\alpha\}\big)$ is the strength of the i-th wave in the solution of the Riemann problem with data $u(x_\alpha-)$, $u(x_\alpha+)$. Call μ_i^+, μ_i^- the positive and negative parts of the signed measure μ_i, so that

$$\mu_i = \mu_i^+ - \mu_i^-, \qquad |\mu_i| = \mu_i^+ + \mu_i^-.$$

The *total strength of waves* in u is defined as

$$V(u) \doteq \sum_{i=1}^{n} V_i(u), \qquad V_i(u) \doteq |\mu_i|(\mathbb{R}), \tag{1.7}$$

while the *interaction potential* of waves in u is

$$\begin{aligned} Q(u) \doteq & \sum_{i<j} \big(|\mu_j| \times |\mu_i|\big)\Big(\big\{(x,y);\ x<y\big\}\Big) \\ & + \sum_i \big(\mu_i^- \times |\mu_i|\big)\Big(\big\{(x,y);\ x \neq y\big\}\Big). \end{aligned} \tag{1.8}$$

With the above notations, the classical interaction estimates [**G, Sm**] can be stated as follows.

PROPOSITION 1.3 (Bounds on wave strengths). *There exists a constant C_1 such that, for every solution $u = u(t,x)$ of (1.1) with small total variation, obtained as limit of wave-front tracking approximations, the following holds. Let $t = \Lambda_j(x)$, $j = 1,2$ be the equations of two space-like curves in the t-x plane, with $\Lambda_1 \leq \Lambda_2$. Then, calling $u_i(x) \doteq u\big(\Lambda_i(x),\ x\big)$, one has*

$$\begin{aligned} & Q(u_2) \leq Q(u_1), \\ & V(u_2) + C_1 \cdot Q(u_2) \leq V(u_1) + C_1 \cdot Q(u_1) \\ & V_i(u_2) + C_1 \cdot Q(u_2) \leq V_i(u_1) + C_1 \cdot Q(u_1) \qquad i = 1, \ldots, n. \end{aligned} \tag{1.9}$$

It is well known that the estimates (1.9) actually hold not only for exact solutions but also for approximate solutions constructed by various algorithms [**B2, G, R**]. For convenience, at various stages of this paper we shall work with slightly different definitions of the interaction potential $Q(u)$. Indeed, one may consider two waves of the same family as being always approaching, regardless of their sign, or as being never approaching. This second definition is useful in connection with systems where shock and rarefaction curves coincide. In all cases, we will make sure that the basic interaction estimates (1.9) remain valid.

As proved in [**B5**], the trajectories of a Standard Riemann Semigroup can be characterized by a set of local integral estimates. Two types of local approximate solutions for (1.1) are considered. One is derived from the self-similar solution of a Riemann problem, the other is obtained by "freezing" the coefficients of the corresponding quasilinear hyperbolic system in a neighborhood of a given point.

Let $u\colon [0,T] \times \mathbb{R} \mapsto \mathbb{R}^n$ be a locally integrable function, and fix any point (τ, ξ) in the domain of u. Assuming $u(\tau, \cdot) \in BV$, consider the limits

$$u^- = \lim_{x \to \xi-} u(\tau, x), \qquad u^+ = \lim_{x \to \xi+} u(\tau, x).$$

Call $\omega = \omega(t,x)$ the self-similar solution of the Riemann problem

$$u_t + \big[F(u)\big]_x = 0, \qquad u(0,x) = \begin{cases} u^- & \text{if } x < 0, \\ u^+ & \text{if } x > 0. \end{cases} \tag{1.10}$$

Let $\hat{\lambda}$ be an upper bound for all characteristic speeds. For $t > \tau$, define

$$U^\sharp_{(u;\tau,\xi)}(t,x) \doteq \begin{cases} \omega(t-\tau,\ x-\xi) & \text{if } |x-\xi| \le \hat{\lambda}(t-\tau), \\ u(\tau,x) & \text{if } |x-\xi| > \hat{\lambda}(t-\tau). \end{cases} \tag{1.11}$$

Observe that the function $t \mapsto U^\sharp_{(u;\tau,\xi)}(t,\cdot)$ is Lipschitz continuous w.r.t. the $\mathbf{L}^1$ distance, and approaches $u(\tau,\cdot)$ as $t \to \tau+$.

Next, call $\widetilde{A} \doteq DF\big(u(\tau,\xi)\big)$ the Jacobian matrix of F computed at the point $u(\tau,\xi)$. For $t > \tau$, define $U^\flat_{(u;\tau,\xi)}(t,x)$ as the solution of the linear hyperbolic Cauchy problem with constant coefficients

$$w_t + \widetilde{A} w_x = 0, \qquad w(\tau,x) = u(\tau,x). \tag{1.12}$$

In the following, by Tot.Var.$\{u(\tau);\ I\}$ we denote the total variation of the function $u(\tau,\cdot)$ over the set I.

PROPOSITION 1.4 (Local integral estimates). *Let $u(t,\cdot) = S_t\bar{u}$ be any semigroup trajectory. Then, for some constant C_2, at each point (τ,ξ) one has*

$$\begin{aligned} &\frac{1}{\eta}\int_{\xi-\rho+\eta\hat{\lambda}}^{\xi+\rho-\eta\hat{\lambda}} \left| u(\tau+\eta,x) - U^\sharp_{(u;\tau,\xi)}(\tau+\eta,x)\right| dx \\ &\qquad \le C_2 \cdot \textit{Tot.Var.}\big\{u(\tau);\ ]\xi-\rho,\ \xi[\ \cup\ ]\xi,\ \xi+\rho[\big\}, \end{aligned} \tag{1.13}$$

$$\begin{aligned} &\frac{1}{\eta}\int_{\xi-\rho+\eta\hat{\lambda}\cdot}^{\xi+\rho-\eta\hat{\lambda}} \left| u(\tau+\eta,x) - U^\flat_{(u;\tau,\xi)}(\tau+\eta,x)\right| dx \\ &\qquad \le C_2 \cdot \Big(\textit{Tot.Var.}\big\{u(\tau);\ ]\xi-\rho,\ \xi+\rho[\big\}\Big)^2, \end{aligned} \tag{1.14}$$

for every $\rho,\eta > 0$ sufficiently small. Viceversa, let $u\colon [0,T] \mapsto \mathbf{L}^1$ be a continuous map taking values inside the domain $\mathcal{D}$ of the semigroup. If the bounds (1.13), (1.14) hold for all $\xi \in \mathbb{R}$ and all but countably many times $\tau \in [0,T]$, then u coincides with a semigroup trajectory.

Following [**B5**], a continuous function $u\colon [0,T] \mapsto \mathbf{L}^1$ will be called a *Viscosity Solution* of (1.1) if the inequalities (1.13), (1.14) hold at every (τ,ξ).

To motivate the decay estimates for waves of genuinely nonlinear families, we consider first the scalar case, assuming $F'' > \kappa > 0$. In this case, if $u = u(t,x)$ is a smooth solution of (1.1) defined for $t \ge \tau$, one has

$$(u_x)_t + F'(u)(u_x)_x = -F''(u)u_x^2. \tag{1.15}$$

Integrating (1.15) along characteristics, one obtains the pointwise estimate [**O, Sm**]

$$u_x(t,x) \le \frac{1}{\kappa(t-\tau)}. \tag{1.16}$$

In the vector-valued case, introduce the gradient components

$$u_x^i \doteq \langle l_i(u),\ u_x\rangle. \tag{1.17}$$

By (1.4), if $u = u(t,x)$ is a smooth solution of (1.1), one has (see [**B1**], p.412)

$$(u_x^i)_t + \big[\lambda_i(u)\big](u_x^i)_x = -\big[\nabla\lambda_i \cdot r_i(u)\big](u_x^i)^2 + \sum_{j\ne k} G_{ijk}(u)u_x^j u_x^k, \tag{1.18}$$

where $G_{ijk}(u) = \lambda_j(u)\langle l_i(u), [r_j(u), r_k(u)]\rangle$. Assume that the i-th field is genuinely nonlinear, so that $\nabla\lambda_i \cdot r_i(u) \geq \kappa > 0$, and let $u = u(t,x)$ be a smooth solution defined for $t \geq \tau$. If $G_{ijk} \equiv 0$ for all i, j, k, then the gradient component u^i_x would clearly satisfy an estimate of the form (1.16). In general, (1.16) may fail because of the last term on the right hand side of (1.18). Observe that this summation essentially depends on the (instantaneous) amount of wave interaction. This suggests that the amount by which (1.16) fails, measured by

$$\int_{\{u^i_x > \eta\}} u^i_x(t,x)\, dx \tag{1.19}$$

for any $\eta > 1/\kappa(t-\tau)$, can be estimated in terms of the total amount of interaction taking place during the interval $[\tau, t]$. This quantity, in turn, can be bounded by $Q(\tau) - Q(t)$, i.e. by the decay in the wave interaction potential. To state the result in the most general case of a BV solution, one more piece of notation is needed. Let $u \in BV$, and let μ_i be the measure determined by the i-waves in u, as in (1.7), (1.8). Call m the Lebesgue measure on $\mathbb{R}$ and split $\mu_i = \mu_i^s + \mu_i^{ac}$ according to its singular and absolutely continuous part. For any $\eta > 0$, define

$$V_i^{\eta+}(u) \doteq \mu_i^{s,+}(\mathbb{R}) + \mu_i^{ac}\left(\left\{x;\ \frac{d\mu_i^{ac}}{dm}(x) > \eta\right\}\right), \tag{1.20}$$

where $\mu_i^{s,+}$ denotes the positive part of μ_i^s. Observe that (1.20) coincides with (1.19) if u is Lipschitz continuous. With the above notation, one has

PROPOSITION 1.5 (Decay Estimates). *Assume that the i-th characteristic field is genuinely nonlinear. Then there exist constants $C_3, \kappa > 0$ such that any solution $u(t,\cdot) = S_t\bar{u}$ of (1.1) satisfies*

$$V_i^{\eta+}(u(t,\cdot)) \leq C_3\big[Q(\tau) - Q(t)\big]\left(1 - \frac{1}{\eta\kappa(t-\tau)}\right)^{-1} \tag{1.21}$$

for every $t > \tau \geq 0$ and $\eta > 1/\kappa(t-\tau)$.

A proof of (1.21) was first derived in [**B-C3**], for weak solutions obtained as limits of wave-front tracking approximations. In Section 7 of this paper we will show that similar estimates hold for our piecewise Lipschitz approximate solutions as well.

Towards a proof of Theorem 1.1, the basic strategy for obtaining a Lipschitz semigroup of solutions of (1.1) is to construct suitable approximate solutions, carefully controlling how their distance varies in time. We recall that, for a scalar conservation law, the entropic solutions constitute a contractive semigroup in $\mathbf{L}^1$ [**C, K**]. Indeed, any two solutions u, u' can be directly compared, showing that the distance

$$\big\|u(t,\cdot) - u'(t,\cdot)\big\|_{\mathbf{L}^1} \tag{1.22}$$

is a non-increasing function of time. This is definitely not true for systems [**T2**]. In the present case, following [**B1**], we consider a one-parameter family of (suitably regular) solutions u^θ, $\theta \in [0,1]$, with $u^0 = u$, $u^1 = u'$. For each t, the distance

(1.22) is clearly bounded by the $\mathbf{L}^1$-length of the path $\gamma_t : \theta \mapsto u^\theta(t,\cdot)$, defined by

$$\|\gamma_t\|_{\mathbf{L}^1} \doteq \sup\left\{ \sum_{j=1}^{\nu} \left\| u^{\theta_j}(t,\cdot) - u^{\theta_{j-1}}(t,\cdot) \right\|_{\mathbf{L}^1},\ 0 = \theta_0 < \theta_1 < \cdots < \theta_\nu = 1,\ \nu \geq 1 \right\}.$$

Therefore, if we show that this length satisfies

$$\|\gamma_t\|_{\mathbf{L}^1} \leq L \cdot \|\gamma_0\|_{\mathbf{L}^1} \tag{1.23}$$

for every curve γ_0 joining $u(0)$ with $u'(0)$, it will follow that

$$\left\| u(t) - u'(t) \right\|_{\mathbf{L}^1} \leq L \cdot \left\| u(0) - u'(0) \right\|_{\mathbf{L}^1} \qquad \forall t \geq 0. \tag{1.24}$$

If the path γ is suitably regular, its length can be computed by integrating the norm of a tangent vector:

$$\|\gamma\|_{\mathbf{L}^1} = \int_0^1 \left\| \frac{d\gamma(\theta)}{d\theta} \right\|_{\mathbf{L}^1} d\theta. \tag{1.25}$$

An estimate of the form (1.24) can thus be obtained by showing that the norm of any tangent vector increases at most by a factor L.

We shall implement the above strategy using paths of piecewise Lipschitz functions. Let $\theta \mapsto u^\theta$ be a one-parameter family of piecewise Lipschitz functions, each u^θ having the same number of jumps, say at the points $y_1^\theta < \cdots < y_N^\theta$. Assume that there exist the functions (Fig. 1.1):

$$v^\theta(x) \doteq \lim_{h \to 0} \frac{u^{\theta+h}(x) - u^\theta(x)}{h} \qquad \text{for a.e. } x \tag{1.26}$$

and the numbers

$$\xi_\alpha^\theta \doteq \lim_{h \to 0} \frac{y_\alpha^{\theta+h} - y_\alpha^\theta}{h} \qquad \alpha = 1, \ldots, N. \tag{1.27}$$

Then, under suitable regularity conditions, the $\mathbf{L}^1$-length of the path $\gamma : \theta \mapsto u^\theta$ is computed by

$$\|\gamma\|_{\mathbf{L}^1} = \int_0^1 \|v^\theta\|_{\mathbf{L}^1} d\theta + \sum_{\alpha=1}^{N} \int_0^1 \left| u^\theta(y_\alpha+) - u^\theta(y_\alpha-) \right| |\xi_\alpha^\theta| \, d\theta. \tag{1.28}$$

We stress the fact that, in general, the path $\theta \mapsto u^\theta$ is not differentiable w.r.t. the usual differential structure of $\mathbf{L}^1$. Indeed, if the shift rates ξ_α^θ are not equal to zero, as $h \to 0$ the ratio $[u^{\theta+h} - u^\theta]/h$ does not converge to any limit in $\mathbf{L}^1$. In order to correctly measure the length of a path γ, it is essential to work with a class of "generalized tangent vectors" of the form $(v, \xi) \in T_u \doteq \mathbf{L}^1(\mathbb{R};\ \mathbb{R}^n) \times \mathbb{R}^N$. Observe that the tangent space T_u actually depends on the function u, through the number of points of discontinuity.

Next, assume that each function $u^\theta(t,\cdot)$ is a solution of the system of conservation laws (1.1). A set of linearized evolution equations for the corresponding tangent vectors (v^θ, ξ^θ) was derived in [**B-M1**]. To write down these equations, some notation must be introduced. Let $A(u)$ be the Jacobian matrix of f at u and

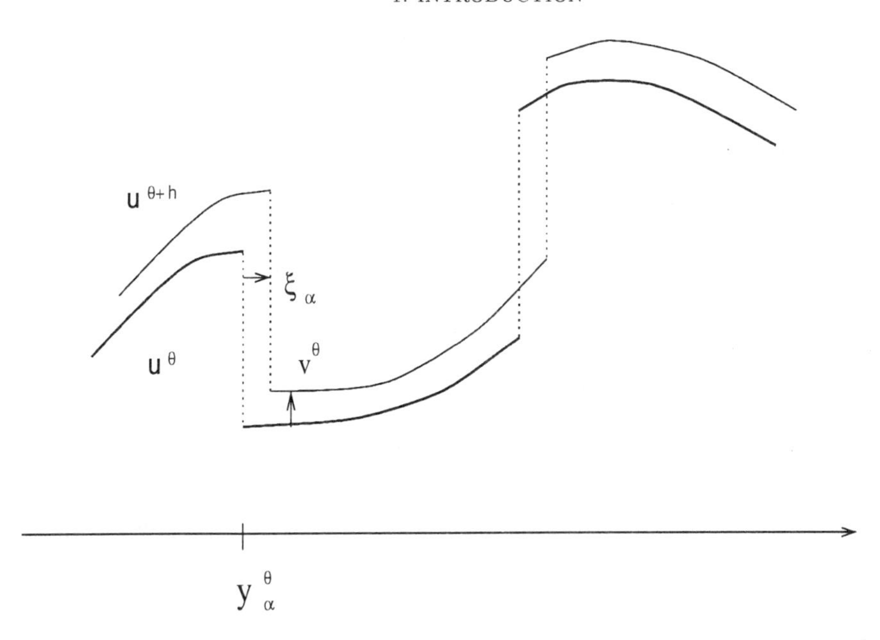

FIGURE 1.1

call $\lambda_i(u), l_i(u), r_i(u)$ respectively its eigenvalues and left and right eigenvectors. For $u, u' \in \mathbb{R}^n$, define the averaged matrix

$$A(u,u') \doteq \int_0^1 A\big(\theta u + (1-\theta)u'\big)d\theta \tag{1.29}$$

with eigenvalues $\lambda_i(u,u')$, and choose right and left eigenvectors $r_i(u,u')$, $l_i(u,u')$ of $A(u,u')$ according to (1.6). The differential of λ_i at (u,u') is written $D\lambda_i(u,u')$. We thus have

$$D\lambda_i(u,u')\cdot(v,v') = \lim_{\varepsilon\to 0} \varepsilon^{-1}\big[\lambda_i(u+\varepsilon v, u'+\varepsilon v') - \lambda_i(u,u')\big]. \tag{1.30}$$

The same notation will be used for the differentials of the eigenvectors r_i and l_i.

Let $u = u(t,x)$ be a piecewise Lipschitz continuous weak solution of (1.1). At almost every point (t,x), the function u thus satisfies the quasilinear system (1.4) while, along the shock lines $x = y_\alpha(t)$, the Rankine-Hugoniot equations hold:

$$\dot y_\alpha[u^+ - u^-] = \big[f(u^+) - f(u^-)\big], \tag{1.31}$$

with $u^+ = u(t, y_\alpha+)$, $u^- = u(t, y_\alpha-)$. If the jump at y_α occurs in the k_α-th characteristic family, this implies that $u^+ - u^-$ is a right eigenvector of the matrix $A(u^+, u^-)$, with corresponding eigenvalue

$$\dot y_\alpha = \lambda_{k_\alpha}(u^+, u^-). \tag{1.32}$$

The eigenvector condition can also be written as

$$\big\langle l_i(u^+, u^-),\ u^+ - u^-\big\rangle = 0 \qquad \forall i \neq k_\alpha. \tag{1.33}$$

A system of linearized evolution equations for the generalized tangent vector (v, ξ) can now be derived from (1.4), (1.32) and (1.33). Namely

$$v_t + A(u)v_x + [DA(u)\cdot v]u_x = 0 \tag{1.34}$$

outside the lines of discontinuity, together with the conditions

$$\begin{aligned}(1.35)\qquad &\Big\langle Dl_i(u^+,u^-)\cdot(\xi_\alpha u_x^+ + v^+,\ \xi_\alpha u_x^- + v^-),\ u^+ - u^-\Big\rangle\\ &\qquad + \Big\langle l_i(u^+,u^-),\ \xi_\alpha u_x^+ + v^+ - \xi_\alpha u_x^- - v^-\Big\rangle = 0 \qquad \forall i \neq k_\alpha,\end{aligned}$$

$$(1.36)\qquad \dot\xi_\alpha = D\lambda_{k_\alpha}(u^+,u^-)\cdot\big(\xi_\alpha u_x^+ + v^+,\ \xi_\alpha u_x^- + v^-\big),$$

on each line $x = y_\alpha(t)$ where u has a shock (or contact discontinuity) in the k_α-th characteristic field. Observe that the equations (1.34) are formally derived from (1.4), replacing u by $u + \varepsilon v$ and differentiating w.r.t. ε. Similarly, the equations (1.35), (1.36) are obtained from (1.33), (1.32) replacing the right and left limits u^+, u^- by $u^+ + \varepsilon v^+ + \varepsilon u_x^+\xi_\alpha$ and $u^- + \varepsilon v^- + \varepsilon u_x^-\xi_\alpha$ respectively, and differentiating w.r.t. ε. For a rigorous derivation of (1.34)–(1.36) see **[B-M1]**.

According to (1.28), the standard $\mathbf{L}^1$-length of a path γ can be computed by integrating the norm of its tangent vector, defined as

$$(1.37)\qquad \big\|(v,\xi)\big\|_{\mathbf{L}^1} \doteq \|v\|_{\mathbf{L}^1} + \sum_\alpha \big|u(y_\alpha+) - u(y_\alpha-)\big|\,|\xi_\alpha|.$$

EXAMPLE 1.6. In the scalar case, it is well known that the conservation law (1.1) generates a contractive semigroup in $\mathbf{L}^1$. In particular, the $\mathbf{L}^1$-length of a path of solutions does not increase in time, and the same holds for the norm (1.37) of a tangent vector. It is an instructive exercise to carry out the computations in this simple case. Let F be convex and let $u = u(t,x)$ be a piecewise Lipschitz solution of (1.1), with jumps at the points $x = y_\alpha(t)$, $\alpha = 1,\ldots,N$. Calling $\lambda(u) \doteq DF(u)$ the characteristic speed, (1.4) takes the form

$$(1.38)\qquad u_t + \lambda(u)u_x = 0.$$

At points of jump, the Rankine-Hugoniot and the entropy conditions yield

$$(1.39)\qquad \dot y_\alpha = \frac{1}{|\sigma_\alpha|}\int_{\sigma_\alpha}^0 \lambda\big(u(y_\alpha-)+s\big)\,ds = \frac{F\big(u(y_\alpha+)\big) - F\big(u(y_\alpha-)\big)}{u(y_\alpha+) - u(y_\alpha-)},$$

$$(1.40)\qquad \sigma_\alpha = u(y_\alpha+) - u(y_\alpha-) < 0.$$

The linearized evolution equations (1.34)–(1.36) for a generalized tangent vector (v,ξ) take the form

$$(1.41)\qquad v_t + \big[\lambda(u)v\big]_x = 0,$$

$$(1.42)\qquad \dot\xi_\alpha = D\lambda(u^-,u^+)\cdot\big(v^- + \xi_\alpha u_x^-,\ v^+ + \xi_\alpha u_x^+\big) \qquad \alpha = 1,\ldots,N.$$

Here and in the following, we use the shorter notation $u^\pm \doteq u(y_\alpha\pm)$, and similarly for $v^\pm, u_x^\pm$. Observing that

$$(1.43)\qquad \dot\sigma_\alpha = \big(\lambda(u^-) - \dot y_\alpha\big)u_x^- + \big(\dot y_\alpha - \lambda(u^+)\big)u_x^+, \qquad \sigma_\alpha = u^+ - u^- < 0,$$

$$
\begin{aligned}
D\lambda(u^-,u^+)\cdot(\eta^-,\eta^+) &\doteq \frac{d}{d\varepsilon}\left[\frac{F(u^-+\varepsilon\eta^-)-F(u^++\varepsilon\eta^+)}{(u^-+\varepsilon\eta^-)-(u^++\varepsilon\eta^+)}\right]_{\varepsilon=0} \\
&= \frac{\lambda(u^-)\eta^- - \lambda(u^+)\eta^+}{u^- - u^+} - \frac{f(u^-)-f(u^+)}{u^- - u^+}\cdot\frac{\eta^- - \eta^+}{u^- - u^+} \\
&= \big(\lambda(u^-)-\dot y\big)\frac{\eta^-}{|\sigma_\alpha|} + \big(\dot y_\alpha - \lambda(u^+)\big)\frac{\eta^+}{|\sigma_\alpha|},
\end{aligned} \tag{1.44}
$$

from (1.41), (1.42) we obtain

$$
\begin{aligned}
&\frac{d}{dt}\Big\{\int_{-\infty}^{\infty}\big|v(t,x)\big|\,dx + \sum_{\alpha=1}^{N}|\xi_\alpha||\sigma_\alpha|\Big\} \\
&\quad = -\Big\{\sum_\alpha \big(\lambda(u^-)-\dot y_\alpha\big)|v^-| + \sum_\alpha\big(\dot y_\alpha - \lambda(u^+)\big)|v^+|\Big\} \\
&\qquad + \sum_\alpha D\lambda(u^-,u^+)\cdot\big(v^- + \xi_\alpha u_x^-,\ v^+ + \xi_\alpha u_x^+\big)(\text{sign }\xi_\alpha)|\sigma_\alpha| \\
&\qquad + \sum_\alpha |\xi_\alpha|\Big[\big(\lambda(u^-)-\dot y_\alpha\big)u_x^- - \big(\dot y_\alpha - \lambda(u^+)\big)\Big] \\
&\quad \le 0.
\end{aligned} \tag{1.45}
$$

In the case of systems, on the other hand, the norm (1.37) may well increase along solutions of (1.34)–(1.36). In [**B4**], however, the following was proved.

PROPOSITION 1.7. *There exists a set $\mathcal{U}$, containing all piecewise Lipschitz functions with sufficiently small total variation, and a family of weighted norms $\|\cdot\|_u$, defined for $u \in \mathcal{U}$, with the following properties.*

(i) *If $u = u(t,x)$ is a piecewise Lipschitz continuous solution of (1.1) with $u(t) \in \mathcal{U}$, and if the pair $\big(v(t),\xi(t)\big)$ is any solution of the corresponding linearized system, then the norm $\big\|(v(t),\xi(t))\big\|_{u(t)}$ is a non-increasing function of time, even at times where two shocks interact.*

(ii) *There exists a constant L such that, for all $u \in \mathcal{U}$ and $(v,\xi) \in T_u$, one has*

$$
\big\|(v,\xi)\big\|_{\mathbf{L}^1} \le \big\|(v,\xi)\big\|_u \le L\cdot\big\|(v,\xi)\big\|_{\mathbf{L}^1}. \tag{1.46}
$$

In turn, the Riemann metric $\|\cdot\|_u$ determines a weighted distance $d_\star$ on $\mathcal{U}$. Roughly speaking, $d_\star(u,u')$ is the infimum of the weighted length of paths joining u with u'. A more careful construction goes as follows.

DEFINITION 1.8. We say that a continuous map $\gamma\colon \theta \mapsto u^\theta \doteq \gamma(\theta)$ from an open interval J into $\mathbf{L}^1_{loc}$ is a *Regular Path* (RP) if the following holds. For $\theta \in J$, all functions u^θ are piecewise Lipschitz continuous, with the same number of jumps, say at $x_1^\theta < \cdots < x_N^\theta$, and the same Lipschitz constant outside these points of jump. They all coincide outside some fixed interval $[-M,M]$. Moreover, the function $\theta \mapsto u_x^\theta$ is continuous from J into $\mathbf{L}^1$. The map $\theta \mapsto u^\theta$ admits a generalized tangent vector $D\gamma(\theta) = (v^\theta,\xi^\theta) \in T_{\gamma(\theta)} = \mathbf{L}^1(\mathbb{R};\mathbb{R}^n)\times\mathbb{R}^N$, continuously depending on θ.

DEFINITION 1.9. A continuous map $\gamma\colon [a,b] \mapsto \mathbf{L}^1$ is a *Piecewise Regular Path* (PRP) if there exist finitely many values $a = \theta_0 < \theta_1 < \cdots < \theta_N = b$ such that the restriction of γ to each open subinterval $J_\ell \doteq]\theta_{\ell-1},\ \theta_\ell[$ is a regular path.

Given any two piecewise Lipschitz continuous functions $u, u' \in \mathcal{U}$, call $\Sigma_{u,u'}$ the family of all regular paths $\gamma\colon [0,1] \mapsto \mathcal{U}$ with $\gamma(0) = u$, $\gamma(1) = u'$. The weighted length of a path $\gamma \in \Sigma_{u,u'}$ is then defined, using the notation of Proposition 1.7, as

$$\|\gamma\|_\star \doteq \int_0^1 \big\|D\gamma(\theta)\big\|_{\gamma(\theta)} d\theta, \tag{1.47}$$

while the Riemannian distance between u and u' is given by

$$d_\star(u, u') \doteq \inf \big\{\|\gamma\|_\star; \quad \gamma \in \Sigma_{u,u'}\big\}. \tag{1.48}$$

As proved in [**B-B**], the weighted length (1.47) is lower semicontinuous: if $(\gamma_\nu)_{\nu \geq 0}$ is a sequence of piecewise regular paths such that

$$\lim_{\nu\to\infty} \sup_{\theta\in[0,1]} \big\|\gamma_\nu(\theta) - \gamma_0(\theta)\big\|_{\mathbf{L}^1} = 0,$$

then

$$\|\gamma_0\|_\star \leq \liminf_{\nu\to\infty} \|\gamma_\nu\|_\star.$$

Because of (ii) in Proposition 1.7, the distance $d_\star$ is uniformly equivalent to the standard $\mathbf{L}^1$ distance. Hence, it can be extended by continuity to the $\mathbf{L}^1$ closure of $\mathcal{U}$. Moreover, by (i), the length of every regular path does not increase in time along the flow of (1.1). This suggests that the flow generated by (1.1) should be globally contractive w.r.t. the weighted distance $d_\star$, and hence uniformly Lipschitz continuous w.r.t. the usual $\mathbf{L}^1$ distance.

Unfortunately, a rigorous proof of this fact runs into a major difficulty. Indeed, the estimates on the norm of a tangent vector, and on the length of a path $\theta \mapsto \gamma_t(\theta)$, are valid assuming that all solutions remain piecewise Lipschitz throughout a given interval $[0, T]$. This is not the case in general. Indeed, a piecewise Lipschitz solution may lose its regularity in two ways (Fig. 1.2):

(i) The number of shock fronts may become infinite in finite time, due to repeated shock interactions.

(ii) The Lipschitz constant outside the shocks may become infinite, due to the genuine nonlinearity of some characteristic fields.

We recall that, in the special case where all characteristic fields are linearly degenerate, solutions which are initially smooth remain smooth for all times. In this case, as soon as Proposition 1.7 has been established, the construction of the semigroup is straightforward [**B1**]. The purpose of the present paper is to show that this construction can still be accomplished, for general $n \times n$ genuinely nonlinear systems, with the aid of three technical tools:

(1) A restarting procedure, which replaces a path $\gamma_\tau: \theta \mapsto u^\theta(\tau)$ with a new path $\gamma_{\tau+}$. This is used when some of the functions u^θ are about to lose their regularity and cannot be prolonged further in time.

(2) A slight modification of the Rankine-Hugoniot equations, which forces shock curves to coincide with rarefaction curves, for small amplitudes.

(3) A cyclical concatenation of flows generated by quasilinear systems where $n-1$ characteristic fields are linearly degenerate and only one is genuinely nonlinear. In the limit, this yields the flow generated by a general system, with an arbitrary number of genuinely nonlinear fields.

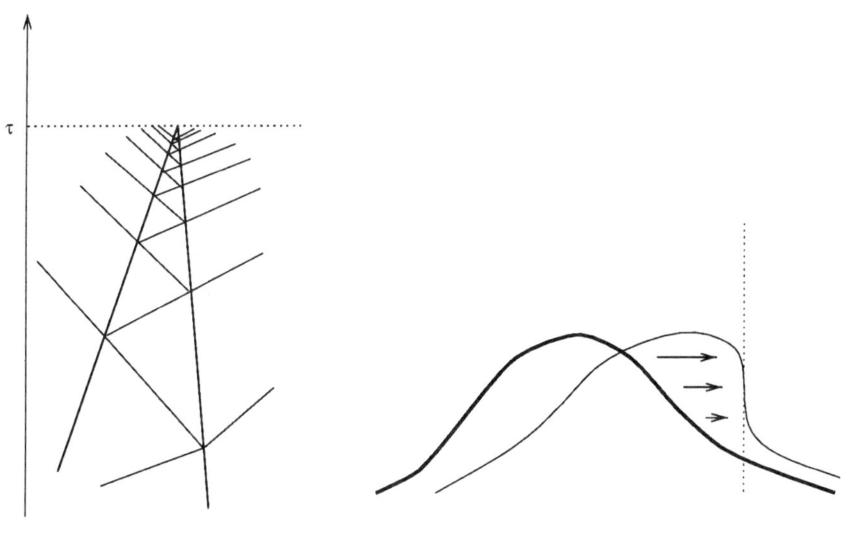

FIGURE 1.2

Restarting procedures, applied to approximate solutions, are well known in the literature. A classical example is the Glimm scheme. Another one, consisting of periodic mollifications, occurs in [**B3**]. We remark that, in our case, the new path $\gamma_{\tau+}$ should satisfy

$$\left\|\gamma_{\tau+}(\theta) - \gamma_\tau(\theta)\right\|_{\mathbf{L}^1} \leq \varepsilon_0 \qquad \forall\theta, \tag{1.49}$$

$$\|\gamma_{\tau+}\|_\star \leq \|\gamma_\tau\|_\star + \varepsilon_0 \tag{1.50}$$

for some $\varepsilon_0 > 0$ suitably small. In other words, the new path must be close to the old one, and its length should be almost the same. In addition, the solutions u^θ should be well defined and remain regular on some interval $[\tau, \tau + \delta]$, with $\delta > 0$ independent of ε_0. Working with exact solutions, it is apparently not possible to meet all these requirements. This is why the approximations **(2)**-**(3)** are used.

The idea of interpolating between shock and rarefaction curves was introduced in [**B-C1**]. In an ε-approximate solution, shocks of strength $|\sigma| \geq 4\varepsilon$ satisfy the Rankine-Hugoniot equations exactly. On the other hand, shocks of strength $\leq 3\varepsilon$ connect a right and a left state lying on the same rarefaction curve. Observing that any solution contains at most finitely many shocks of strength $> 3\varepsilon$, the advantage of this approximation is clear. Indeed, outside a finite number of lines in the t-x plane (which we regard as "free boundaries"), we are now dealing with a system where shock and rarefaction curves coincide.

The cyclical concatenation of flows, a particular kind of flux-splitting method, is the main new technique introduced in the present paper. It provides the key for extending the result in [**B-C1**] to the general $n \times n$ case. The construction goes as follows. Given the system (1.4), fix a state u^* and let $\lambda_1^* < \cdots < \lambda_n^*$ be the eigenvalues of $A(u^*)$. For $h = 1, \ldots, n$, call $A_h(u)$ the matrix with the same eigenvectors $r_1(u), \ldots, r_n(u)$ as $A(u)$, but whose eigenvalues are

$$\lambda_1^*, \ \ldots \ , \ \lambda_{h-1}^*, \ \lambda_h^* + n\big(\lambda_h(u) - \lambda_h^*\big), \ \ldots \ , \ \lambda_n^*. \tag{1.51}$$

Clearly, for each h, the system

$$u_t + A_h(u)u_x = 0 \tag{1.52}$$

has $n-1$ linearly degenerate fields. Call $(t, \bar{u}) \mapsto S_t^h \bar{u}$ the corresponding flow. Given a time step $\Delta t > 0$, we now concatenate the flows of the semigroups $S^1, \ldots, S^n$ cyclically, on subintervals of length $\Delta t/n$. Letting $\Delta t \to 0$, in the limit we obtain the flow determined by (1.4).

CHAPTER 2

Outline of the proof

We collect here the basic steps in the proof of Theorem 1.1. Technical details will be worked out in the remaining sections.

By possibly performing a linear rescaling of time, it is not restrictive to assume that all wave speeds are < 1 in absolute value. Moreover, throughout the main construction we shall assume that all characteristic fields are genuinely nonlinear. When one or more linearly degenerate fields are present, the minor modifications needed in the proof will be described in the last section of the paper.

We begin by defining, for a given $\varepsilon > 0$, a set of approximate Rankine-Hugoniot conditions. These coincide with the usual ones for shocks of strength $|\sigma| \geq 4\varepsilon$. For a given state $u \in \mathbb{R}^n$ and $i = 1, \ldots, n$, denote by

$$\sigma \mapsto S_i(\sigma)(u), \qquad \sigma \mapsto R_i(\sigma)(u),$$

the usual i-shock and i-rarefaction curves through u, parametrized by arclength. As customary, the orientation is chosen so that the i-th characteristic speed is increasing along the curves S_i, R_i. Consider a smooth, non-increasing map φ: $\mathbb{R} \mapsto$

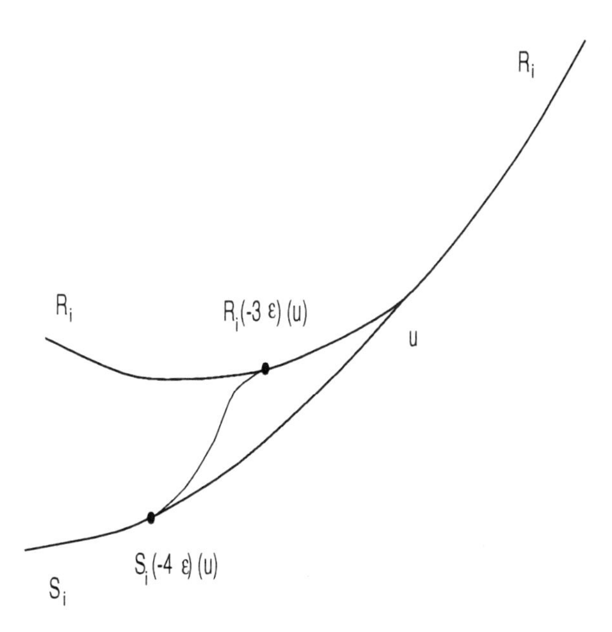

FIGURE 2.1

$[0,1]$ such that

$$\begin{cases} \varphi(\sigma) = 1 & \text{if } \sigma < -4 \\ \varphi(\sigma) = 0 & \text{if } \sigma > -3 \end{cases} \qquad \dot\varphi \in [-2,0]. \tag{2.1}$$

Define the interpolated curves (Fig. 2.1)

$$\Psi_i^\varepsilon(\sigma)(u) \doteq \varphi(\sigma/\varepsilon)\cdot S_i(\sigma)(u^-) + \big(1-\varphi(\sigma/\varepsilon)\big)\cdot R_i(\sigma)(u^-). \tag{2.2}$$

DEFINITION 2.1. We say that the jump (u^-,u^+), located at $x = y(t)$ and travelling with speed $\dot y$, satisfies the *ε-Rankine-Hugoniot conditions* (ε-RH) if, for some $\sigma < 0$ and some $i \in \{1,\dots,n\}$, we have

$$u^+ = \Psi_i^\varepsilon(\sigma)(u^-), \tag{2.3}$$

$$\dot y = \lambda_i^\varepsilon(u^-,u^+) \doteq \varphi(\sigma/\varepsilon)\lambda_i^s + \big(1-\varphi(\sigma/\varepsilon)\big)\lambda_i^r. \tag{2.4}$$

Here λ_i^s is the speed of a true shock connecting u^- with $S_i(\sigma)(u^-)$, while

$$\lambda_i^r \doteq \frac{1}{|\sigma|}\int_\sigma^0 \lambda_i\big(R_i(s)(u^-)\big)\,ds. \tag{2.5}$$

The modified Rankine-Hugoniot conditions introduced above, in turn, determine a new way of approximately solving a Riemann problem. More precisely, consider the initial data

$$u(0,x) = \begin{cases} u^- & \text{if } x < 0, \\ u^+ & \text{if } x > 0. \end{cases} \tag{2.6}$$

We seek a self-similar, piecewise Lipschitz function $\omega = \omega(t,x)$ which satisfies the quasilinear system (1.4) almost everywhere, and the ε-approximate Rankine-Hugoniot conditions along each shock line. The solution to this problem is provided by the following

ε-Riemann Solver: Using the implicit function theorem, determine wave sizes $\sigma_1,\dots,\sigma_n$ such that

$$u^+ = \Psi_n^\varepsilon(\sigma_n)\circ\cdots\circ\Psi_1^\varepsilon(\sigma_1)(u^-). \tag{2.7}$$

Let $\omega_0 = u^-$, $\omega_1 = \Psi_1^\varepsilon(\sigma_1)(\omega_0)$, $\dots$, $\omega_n = \Psi_n^\varepsilon(\sigma_n)(\omega_{n-1}) = u^+$ be the corresponding intermediate states. If $\sigma_i \geq 0$, the states ω_{i-1} and ω_i are connected by a centered rarefaction wave of the i-th family, as usual. If $\sigma_i < 0$, these two states are connected by a single jump, travelling with the speed $\dot y = \lambda_i^\varepsilon(\omega_{i-1},\omega_i)$ defined at (2.4).

DEFINITION 2.2. An $\mathbf{L}^1$–continuous map $u\colon [0,T[\mapsto BV$ is a *Viscosity ε-solution* of (1.1) if at each point (τ,ξ) the inequalities (1.13), (1.14) hold for all $\rho,\eta > 0$ sufficiently small, with $U^\flat$ the solution of (1.12), and

$$U^\sharp_{(u;\tau,\xi)}(t,x) \doteq \begin{cases} \omega^\varepsilon(t-\tau,\ x-\xi) & \text{if } |x-\xi| \leq t-\tau, \\ u(\tau,x) & \text{if } |x-\xi| > t-\tau. \end{cases}$$

Here ω^ε is the ε-solution of the Riemann problem with data $u(\tau,\xi-), u(\tau,\xi+)$. Observe that, by our initial assumption, we can take here $\hat\lambda = 1$ as an upper bound for all wave speeds.

For the most part, our work will be devoted to a proof of

THEOREM 2.3. *With the same assumptions of Theorem 1.1, for some constants* $L, \eta_0 > 0$ *the following holds. For each* $\varepsilon > 0$ *there exists a closed domain* $\mathcal{D}^\varepsilon \subset \mathbf{L}^1(\mathbb{R};\ \mathbb{R}^n)$ *and a continuous semigroup* $S^\varepsilon : \mathcal{D}^\varepsilon \times [0, \infty[\mapsto \mathcal{D}^\varepsilon$ *with the properties:*

(i) *Every function* $\bar{u} \in \mathbf{L}^1$ *with* $Tot.Var.(\bar{u}) \leq \eta_0$ *lies in* $\mathcal{D}^\varepsilon$.

(ii) *For all* $\bar{u}, \bar{v} \in \mathcal{D}^\varepsilon$, $t, s \geq 0$ *one has* $\left\|S_t^\varepsilon \bar{u} - S_s^\varepsilon \bar{v}\right\|_{\mathbf{L}^1} \leq L \cdot \left(|t-s| + \|\bar{u} - \bar{v}\|_{\mathbf{L}^1}\right)$.

(iii) *If* $\bar{u} \in \mathcal{D}^\varepsilon$ *is piecewise constant, then for* $t > 0$ *sufficiently small the function* $u(t, \cdot) = S_t^\varepsilon \bar{u}$ *coincides with the solution of (1.1)-(1.2) obtained by piecing together the solutions of the corresponding Riemann problems determined by the* ε*-Riemann Solver.*

As soon as Theorem 2.3 is established, letting $\varepsilon \to 0$ it will be an easy matter to show that the semigroups S^ε converge to a unique semigroup S, having all the properties stated in Theorem 1.1. Throughout the following, we thus fix some $\varepsilon > 0$ and concentrate on the construction of the semigroup S^ε, with constants $L, \eta_0 > 0$ independent of ε. This requires several steps.

STEP 1. We begin by showing that a Lipschitz semigroup exists, whose domain contains all suitably small perturbations of a Riemann data.

PROPOSITION 2.4. *There exists a neighborhood of the origin* $\Omega_0 \subset \mathbb{R}^n$ *and positive constants* $L, \eta = \eta(\varepsilon) > 0$ *for which the following holds. For every* $u^-, u^+ \in \Omega_0$*, there exists a closed domain* $\mathcal{D} = \mathcal{D}_{(u^-,u^+)}$ *of the form*

(2.8)
$$\mathcal{D} \doteq \left\{ u \in BV;\ \int_{-\infty}^{0} \left|u(x) - u^-\right| dx + \int_{0}^{\infty} \left|u(x) - u^+\right| dx < \infty,\ Q(u) \leq \eta \right\},$$

and a Lipschitz semigroup S *on* $\mathcal{D}$ *satisfying (ii) and (iii) in Theorem 2.3 (with* $\mathcal{D}^\varepsilon$ *replaced by* $\mathcal{D}$*). Such a semigroup is unique, up to the domain. Its trajectories are Viscosity* ε*-solutions of (1.1).*

It should be noted that Proposition 2.4 is much weaker than Theorem 2.3. Indeed, in the theorem the constant η_0 is independent of ε. This is essential, since we eventually need to consider the limit as $\varepsilon \to 0$. In Proposition 2.4, however, we allow ourselves to choose η small depending on ε. In particular, by taking $\eta << \varepsilon^2$, we can assume that every function $u \in \mathcal{D}_{(u^-,u^+)}$ has the same number of large shocks of strength $|\sigma| > \varepsilon$ (i.e. the same "qualitative structure") as the solution of the Riemann problem with data u^-, u^+.

The construction of the semigroup S relies on the ideas described in the Introduction. For sake of clarity, we first describe the constructive procedure in the special case where the solution of the Riemann problem (2.6) does not contain any shock of strength $\geq 2\varepsilon$. In this case, by choosing $\eta > 0$ sufficiently small, we can assume that all functions u in the set $\mathcal{D} = \mathcal{D}_{(u^-,u^+)}$ in (2.8) contain shocks only of strength $< 3\varepsilon$. Because of the definitions (2.1)-(2.2), inside the positively invariant domain $\mathcal{D}$ the time evolution is thus determined by a system where shock and rarefaction curves coincide.

On a given interval $[0, T]$, our approximations will be piecewise Lipschitz solutions of a quasilinear system, except for a finite number of times $0 = t_0 < t_1 < \cdots < t_\nu < T$ where a restarting procedure is used. Namely, to prevent a loss of regularity, at suitable times t_ℓ we shall replace $u(t_\ell-,\cdot)$ with a "nicer" function $u(t_\ell+,\cdot)$.

First of all, we choose a time step $\Delta t = \delta_1 > 0$ and partition the time axis into intervals of the form

$$I_{m,h} \doteq \left[\tau_{m,h-1},\ \tau_{m,h}\right[\ \doteq\ \left[\Big(m+\frac{h-1}{n}\Big)\delta_1,\ \Big(m+\frac{h}{n}\Big)\delta_1\right[, \tag{2.9}$$

with $m \in \mathbb{N}$, $h \in \{1,\dots,n\}$. Moreover, we fix a constant state u^* and call $\lambda_1^* < \cdots < \lambda_n^*$ the eigenvalues of the matrix $A(u^*)$. For $h = 1,\dots,n$, let $A_h(u)$ be the matrix with the same eigenvectors as $A(u)$, but whose eigenvalues are

$$\lambda_1^*,\ \dots,\lambda_{h-1}^*,\ \lambda_h^* + n\big(\lambda_h(u) - \lambda_h^*\big),\ \dots,\lambda_n^*. \tag{2.10}$$

In other words, for every $\mathbf{v} \in \mathbb{R}^n$,

$$A_h(u)\mathbf{v} = n\big(\lambda_h(u) - \lambda_h^*\big)\big(l_h(u)\cdot\mathbf{v}\big)r_h(u) + \sum_{i=1}^{n}\lambda_i^*\big(l_i(u)\cdot\mathbf{v}\big)r_i(u).$$

On each subinterval $I_{m,h}$, our approximation u will have the following properties. Each $u(t,\cdot)$ is piecewise Lipschitz continuous and satisfies

$$u_t + A_h(u)u_x = 0 \tag{2.11}$$

a. e. outside the jumps. At every point $x_\alpha(t)$ where $u(t,\cdot)$ is discontinuous, the jump occurs always in the h-characteristic family and the following approximate Rankine-Hugoniot conditions hold:

$$u(t,x_\alpha+) = R_h(\sigma_\alpha)\big(u(t,x_\alpha-)\big), \tag{2.12}$$

$$\dot{x}_\alpha = \frac{1}{|\sigma_\alpha|}\int_{\sigma_\alpha}^{0}\lambda_h\Big(R_h(s)\big(u(t,x_\alpha-)\big)\Big)\,ds, \tag{2.13}$$

for some $\sigma_\alpha < 0$. As usual, $\sigma \mapsto R_h(\sigma)(u)$ describes the h-rarefaction curve through the state u.

For a given piecewise Lipschitz initial data, the local existence and uniqueness of a piecewise Lipschitz solution to (2.11)–(2.13) follows from the standard theory of quasilinear hyperbolic equations. Let us briefly examine for how long this solution can retain its piecewise Lipschitz regularity. To fix the ideas, assume that, at some time $\tau \in I_{m,h}$, the gradient components $u_x^i = l_i(u)\cdot u_x$ of $u(\tau,\cdot)$ satisfy the bounds (for a. e. $x \in \mathbb{R}$)

$$|u_x^i| \le M \qquad i = 1,\dots,n, \tag{2.14}$$

$$u_x^h \ge -1. \tag{2.15}$$

(i) Whenever two or more h-shocks interact, they simply join together forming a single h-shock, without generating outgoing waves of any other family. Indeed, by construction all shocks in u belong to the h-family. Moreover, according to our modified dynamics, shock and rarefaction curves coincide. As a consequence, the number of shock fronts can only decrease in time.

(ii) For $i \ne h$, the i-th eigenvalue of the matrix $A_h(u)$ is constantly equal to λ_i^*. Hence, for the system (2.11), the i-th family is linearly degenerate. As a consequence, all gradient components u_x^i, $i \ne h$, remain uniformly bounded. To

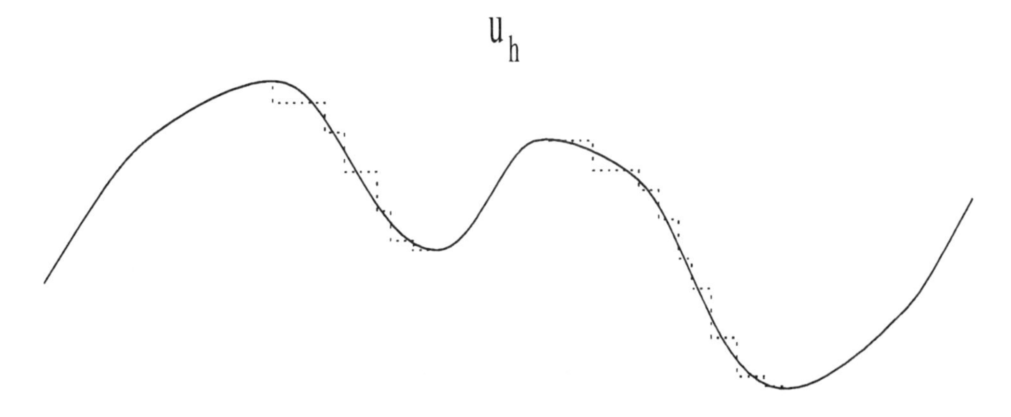

FIGURE 2.2

estimate the growth in the (genuinely nonlinear) component u_x^h, let $x = x(t)$ be an h-characteristic line, so that

$$\dot{x}(t) = \lambda_h^* + n\Big(\lambda_h\big(u(t,x)\big) - \lambda_h^*\Big).$$

Differentiating along this line, from (2.11) we obtain an equation of the form

$$\frac{d}{dt}u_x^h\big(t,x(t)\big) = -n\big[\nabla\lambda_h \cdot r_h(u)\big]\cdot(u_x^h)^2 + \sum_{i\neq j} G_{hij}(u)u_x^i u_x^j, \tag{2.16}$$

where the functions G_{hij} are uniformly bounded. Due to the squared term on the right-hand side, the component u_x^h may well approach $-\infty$ in finite time. However, by (2.16) and (2.15), this component can be bounded from below in terms of the solution to the O. D. E.

$$\dot{Z} = -aZ^2 + bZ - c, \qquad\qquad Z(0) = -1, \tag{2.17}$$

where the positive constants a, b, c depend only on the coefficients G_{hij} and on the uniform bounds already available on the other components u_x^i, $i \neq h$. This provides a lower bound on the time when the solution u of (2.11)-(2.13) may lose its piecewise Lipschitz regularity.

Since the regularity of solutions is not preserved globally in time, the construction of piecewise Lipschitz approximate solutions on a given interval $[0, T]$ must involve some restarting procedures. These restartings are of two types.

(1) At time $t = 0$ we approximate the initial data $\bar{u}$ with a piecewise Lipschitz function $u(0+, \cdot)$ containing only 1-shocks. At each time $t = \tau_{m,h}$ at (2.9), we replace the function $u(t-, \cdot)$ (which contains only h-shocks) with a new function $u(t+, \cdot)$ containing only $(h+1)$-shocks [only 1-shocks in the case $h = n$].

(2) At a time $t \in I_{m,h}$ when $\inf_x u_x^h(t,x)$ is getting close to $-\infty$, we replace $u(t-, \cdot)$ with a new function $u(t+, \cdot)$, still containing only h-shocks, which satisfies (2.15). This can always be accomplished by inserting several small downward jumps in regions where u_x^h is large and negative (Fig. 2.2).
In all cases, at each restarting time t, we require that the distance $\big\|u(t+, \cdot) - u(t-, \cdot)\big\|_{\mathbf{L}^1}$ be suitably small. Moreover, we make sure that the total strength

of waves and the interaction potential are changed very little by the restarting procedure.

By the analysis in **(i)-(ii)**, the construction of an approximate solution on a given interval $[0, T]$ can be accomplished with a finite number of restartings. Indeed, on each subinterval $I_{m,h}$ we can derive an a-priori bound of the form (2.14) on every linearly degenerate component u^i_x, $i \neq h$. Hence, by (2.15) and (2.16), the difference $t_{\ell+1} - t_\ell$ between two restarting times is bounded from below by the length of the interval where the solution of (2.17) is defined.

We can now repeat the above construction with different values of the time step δ_1, letting $\delta_1 \to 0$, and obtain a sequence of approximate solutions $(v_\nu)_{\nu\geq 1}$. By possibly taking a subsequence, a compactness argument yields some function $u = \lim_\nu u_\nu$. We claim that this limit provides an ε-solution to (1.1)-(1.2).

Intuitively, this is seen as follows. Consider any wave, say of the k-th family. In an exact solution, this wave should travel with speed $\lambda_k(u)$. On the other hand, in an approximate solution with time step $\Delta t = \delta_1$, by (2.10) such a wave will travel with speed

$$\dot{x} = \begin{cases} \lambda_k^* & \text{if } t \notin \bigcup_{m\geq 1} I_{m,k}, \\ \lambda_k^* + n\big(\lambda_k(u) - \lambda_k^*\big) & \text{if } t \in \bigcup_{m\geq 1} I_{m,k}. \end{cases}$$

Therefore, on any interval of the form

$$\big[m\delta_1,\ (m+1)\delta_1\big[\ =\ I_{m,1} \cup \cdots \cup I_{m,n},$$

the average speed of a k-wave coincides with $\lambda_k(u)$. Letting $\delta_1 \to 0$ we thus obtain an exact solution.

Of course, one could construct approximate solutions by a less sophisticated technique, such as the Glimm scheme or wave-front tracking. The use of piecewise Lipschitz approximations with successive restartings, however, offers a major advantage compared with all other methods. Namely, the distance between any two approximate solutions can now be carefully estimated. This is done as follows (Fig. 2.3).

For a fixed $\delta_1 > 0$, let u, u' be any two approximate solutions, constructed on the interval $[0, T]$ according to the above procedure. Consider any piecewise regular path $\gamma_0 \colon \theta \mapsto u^\theta(0, \cdot)$ joining $u(0, \cdot)$ with $u'(0, \cdot)$. We can then construct a one-parameter family of approximate solutions u^θ within the class of piecewise Lipschitz functions, implementing a restarting procedure at suitable times t_ℓ and carefully controlling the length of the path $\gamma_t \colon \theta \mapsto u^\theta(t, \cdot)$.

At times where all functions u^θ solve the same quasilinear system (2.11) with jump conditions (2.12)-(2.13), the weighted length of the generalized tangent vector $\|du^\theta/d\theta\|_{u^\theta}$ does not increase. Moreover, a careful construction guarantees that, at each time t where a restarting procedure is applied (simultaneously to all functions $u^\theta(t-, \cdot)$), the weighted length of the path γ_t changes only by a very small amount. In the end, we obtain a bound on the length of the path γ_T, and hence on the distance $\big\|u(T, \cdot) - u'(T, \cdot)\big\|_{\mathbf{L}^1}$, based on the length of the path γ_0. Letting $\delta_1 \to 0$, this yields the Lipschitz estimate (ii) in Theorem 2.3.

It is worth observing that, if (1.1) is a special system where all shock and rarefaction curves coincide **[T1]**, then the above construction already provides a

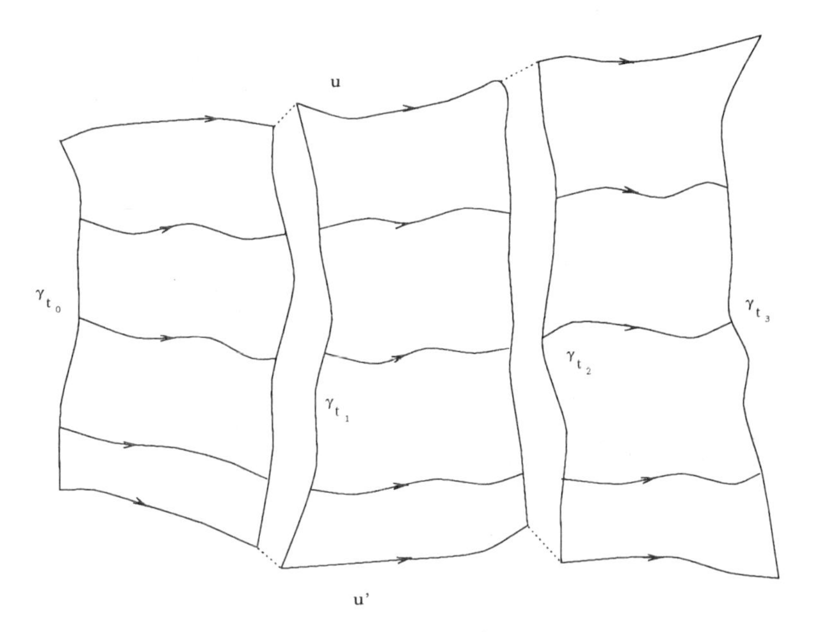

FIGURE 2.3

complete proof of Theorem 1.1. For general systems, however, we still have to describe a proof of Proposition 2.4, in the case where the solution of the Riemann problem (u^-, u^+) contains one or more shocks with strength $|\sigma_i| \geq 2\varepsilon$.

To fix the ideas, assume that the ε-solution of the Riemann problem (1.1)-(2.6) has ν shocks of strength $\geq 2\varepsilon$, say in the families $j_1, \ldots, j_\nu \in \{1, \ldots, n\}$, plus possibly other shocks of smaller strength. Then, if we choose $\eta << \varepsilon^2$ sufficiently small, every function $u \in \mathcal{D}_{(u^-,u^+)}$ will contain exactly one shock of strength $> \varepsilon$ for each of the families $j_1, \ldots, j_\nu$, plus possibly other shocks (of different families) all with strength $< 3\varepsilon$. For any fixed $\delta_1, \delta_2 > 0$, a piecewise Lipschitz continuous approximation $u = u(t,x)$ is constructed by choosing a time step $\Delta t = \delta_1$ and inserting ν "shock layers" of width δ_2 around the large shocks, say located at $y_{j_1}(t) < \cdots < y_{j_\nu}(t)$. For $t \in I_{m,h}$ as in (2.9), the function u will have the following properties.

- At each point y_i, $i \in \{j_1, \ldots, j_\nu\}$, the ε-Rankine-Hugoniot conditions (2.3)-(2.4) hold.
- Inside each shock layer

$$B_i \doteq \big[y_i(t) - \delta_2,\ y_i(t) + \delta_2\big] \qquad i \in \{j_1, \ldots, j_\nu\}, \tag{2.18}$$

  the evolution of u is determined by a quasilinear system where all characteristic speeds are constant (hence all fields are linearly degenerate).
- Outside the shock layers B_i, the function u is a solution of the quasilinear system (2.11), with jumps occurring only in the h-characteristic field. All these jumps have strength $|\sigma| < 3\varepsilon$ and satisfy the ε-Rankine-Hugoniot equations (2.12)-(2.13).

The reason for inserting the shock layers is the following. Within a time interval $I_{m,h}$, if a small h-shock were to hit one of the large shocks, the interaction could produce several outgoing shocks of various families. By subsequent interactions, the total number of shocks could thus approach $+\infty$ within a very short time, and the algorithm would break down. To avoid this, we adopt an additional restarting procedure:

(3) At a time t when an h-shock has penetrated inside one of the shock layers B_i, this shock is replaced by a steep Lipschitz continuous compression wave, before hitting the big shock at $y_i(t)$.

By letting all wave speeds be constant inside B_i, we make sure that this steep compression wave (of the h-family) does not "break", reforming an h-shock almost immediately. Since all wave speeds are bounded, it takes a positive amount of time for an h-wave to penetrate the shock layer B_i of width $\delta_2 > 0$ and hit the shock at $y_i(t)$. Therefore, the length of the interval $[t_\ell,\ t_{\ell+1}]$ between two consecutive restarting times remains uniformly positive. In a finite number of steps, our approximate solution can thus be constructed on the whole interval $[0, T]$.

A detailed construction of approximate solutions in this more general case will be given in Section 3, together with the estimates on the weighted length of a generalized tangent vector. The restarting procedures are carefully analyzed in Section 4. The proof of Proposition 2.4 is then completed in Section 5, showing that, as $\delta_1, \delta_2 \to 0$, our approximate solutions converge to a unique limit, continuously depending on the initial data.

STEP 2. For a given $\varepsilon > 0$, we now construct a continuous semigroup S^ε on a domain $\mathcal{D}^\varepsilon$, satisfying (i) and (iii) in Theorem 2.3.

Observe that, in Step 1, we obtained a family of semigroups with the properties (ii) and (iii) but not (i), because the domains $\mathcal{D}_{(u^-,u^+)}$ would shrink to a point as $\varepsilon \to 0$. On the other hand, the domains $\mathcal{D}^\varepsilon$ which we now consider remain uniformly large, satisfying (i) for some $\eta_0 > 0$ independent of ε. The trajectories of the semigroup S^ε are constructed by a wave-front tracking algorithm. This algorithm is essentially the same as described in [**B-J, B6**], except for the use of the ε-approximate Riemann solver (2.7).

The basic idea in the wave-front tracking method is well known [**D, DP2, B2, R**]. Start with a piecewise constant function $u(0,\cdot)$ close to the initial data $\bar{u}$. At each point of jump, one approximately solves the corresponding Riemann problem within the class of piecewise constant functions. This yields an approximate solution defined up to the first time t_1 where one or more wave-front interactions take place. The new Riemann problem is then solved again within the class of piecewise constant functions, prolonging the solution up to some time t_2 where the second set of interactions takes place, etc. In order to keep finite the total number of wave-fronts, we shall use two distinct procedures for approximately solving a Riemann problem: an accurate method, which possibly introduces several new fronts, and a simplified method, which avoids the introduction of new wave-fronts. The algorithm involves two (strictly positive) parameters: $\bar{\sigma}$, bounding the maximum size of rarefaction fronts, and $\bar{\rho}$, determining which Riemann problems will be accurately solved.

For a given Riemann problem with data u^-, u^+, these two solution methods are described below. By $[\![s]\!]$ we denote here the integer part of s.

Accurate ε-Riemann solver: Let $\omega_0 = u^-$, $\omega_1, \ldots, \omega_n = u^+$ be the states present in the ε-approximate solution of the Riemann problem, as in (2.7). The piecewise constant approximation u is then obtained by replacing each rarefaction wave with a rarefaction fan. More precisely, if $\omega_i = \Psi_i^\varepsilon(\sigma_i)(\omega_{k-1})$ with $\sigma_i > 0$, we divide this jump into $p_i = 1 + [\![\sigma_i/\bar{\sigma}]\!]$ equal parts, inserting the intermediate states

$$\omega_{i,0} = \omega_{i-1},\ \omega_{i,1},\ \ldots,\ \omega_{i,p_i} = \omega_i.$$

Each small jump $(\omega_{i,\ell-1}, \omega_{i,\ell})$ travels with speed $\lambda_i(\omega_{i,\ell})$, i.e. with the characteristic speed of its right state.

Simplified ε-Riemann Solver: Assume that the Riemann problem is determined by the interaction of two waves of distinct families, say $i < j$, with sizes σ_i, σ_j. Call u^l, u^m, u^r the left, middle and right states before the interaction. Clearly, $u^m = \Psi_j^\varepsilon(\sigma_j)(u^l)$, $u^r = \Psi_i^\varepsilon(\sigma_i)(u^m)$. We then solve the Riemann problem in terms of two outgoing wave-fronts of the same families, still with sizes σ_i, σ_j. The solution will thus involve a middle state $\tilde{u}^m = \Psi_i^\varepsilon(\sigma_i)(u^l)$ and a new right state $\tilde{u}^r = \Psi_j^\varepsilon(\sigma_j)(\tilde{u}^m)$. In general, $\tilde{u}^r \neq u^r$. The jump $(\tilde{u}^r, u^r)$ is then propagated along a *non-physical* wave-front, travelling with a fixed speed $\hat{\lambda}$, larger than all characteristic speeds.
In the case where both incoming wave-fronts belong to the same i-th family and have sizes σ_i, σ_i', the Riemann problem is solved by a single outgoing i-wave of size $\sigma_i + \sigma_i'$, together with a non-physical wave-front connecting the states $\Psi_i^\varepsilon(\sigma_i + \sigma_i')(u^l)$ and u^r, travelling with speed $\hat{\lambda}$.
Finally, in the case where a non-physical front hits an i-wave of size σ_i, the Riemann problem is solved in terms of an outgoing i-wave of the same size σ_i, and a non-physical front always travelling with speed $\hat{\lambda}$.

To complete the description of the algorithm, it remains to specify which Riemann solver is used at any given interaction:

- The accurate method is used at time $t = 0$, and at every interaction where the product of the strengths of the incoming waves is $\geq \bar{\rho}$.
- The simplified method is used at every interaction involving a non-physical wave-front, and also at interactions where the product of the strengths of the incoming waves is $< \bar{\rho}$.

In the above, we tacitly assumed that only two wave-fronts interact at any given point. This can always be achieved by an arbitrarily small change in the speed of one of the interacting fronts.

Given any initial data $\bar{u}$ with sufficiently small total variation, consider a sequence of piecewise constant functions such that $\bar{u}_\nu \to \bar{u}$ in $\mathbf{L}^1$. Calling N_ν the number of jumps in $\bar{u}_\nu$, choose the parameter values

$$\bar{\sigma}_\nu \doteq \frac{1}{\nu}, \qquad \bar{\rho}_\nu \doteq e^{-(N_\nu + \nu)}.$$

For each $\nu \geq 1$, using the above algorithm we now construct a piecewise constant approximate solution u_ν, with $u_\nu(0, \cdot) = \bar{u}_\nu$. Each u_ν is defined for all $t \in [0, \infty[$ and has a finite number of lines of discontinuity in the t-x plane. By a straightforward adaptation of the estimates in [**B-J, B6**] one checks that, as $\nu \to \infty$,

(i) The total variation of $u_\nu(t, \cdot)$ remains uniformly small,

(ii) The maximum strength of rarefaction fronts in u_ν approaches zero,
(iii) The total strength of all non-physical waves approaches zero.

By (i), Helly's theorem guarantees the existence of a subsequence which converges to some function u in $\mathbf{L}^1_{loc}$. By (ii) and (iii), u is an ε-solution of (1.1). Moreover, relying on the lengthy construction performed in Step 1, we can show that this limit function u is unique and depends continuously on the initial data.

PROPOSITION 2.5. *Let $\bar u \in \mathbf{L}^1$ have sufficiently small total variation. Then any sequence of approximations u_ν generated by the above wave-front tracking algorithm converges to a unique limit $u = u(t,x)$, continuously depending on the initial data $\bar u$. The map*

$$(t, \bar u) \mapsto u(t, \cdot) \doteq S^\varepsilon_t \bar u$$

is a continuous semigroup. Every trajectory is a Viscosity ε-solution of the corresponding Cauchy problem (1.1), (1.2).

The key step in the proof of Proposition 2.5 is to show that, locally in the t-x plane, the limit solution u coincides with a trajectory of one of the Lipschitz semigroups constructed in Step 1. More precisely, fix any point $(\tau, \bar x)$ and define the one-sided limits $u^- = u(\tau, \bar x-)$, $u^+ = u(\tau, \bar x+)$. Choosing $\rho > 0$ small enough, the truncated function

$$\tilde u(x) = \begin{cases} u(\tau, x) & \text{if } x \in [\bar x - \rho,\ \bar x + \rho], \\ u^- & \text{if } x < \bar x - \rho, \\ u^+ & \text{if } x > \bar x + \rho, \end{cases} \tag{2.19}$$

lies in the domain $\mathcal{D}_{(u^-, u^+)}$ of one of the semigroups S constructed in Step 1. One then has

$$\big(S_{t-\tau} \tilde u\big)(x) = u(t,x) \qquad \forall (t,x) \in \Gamma, \tag{2.20}$$

where Γ is the domain of dependency

$$\Gamma \doteq \big\{(t,x);\ t \geq \tau,\ |x - \bar x| \leq \rho - (t-\tau)\big\}. \tag{2.21}$$

From the uniqueness of the semigroup S, proved as in [**B4**], it thus follows the uniqueness of the limit function u.

STEP 3. All the remaining analysis aims at establishing the uniform Lipschitz continuity of the semigroup S^ε, with a Lipschitz constant independent of ε, thus proving (ii) in Theorem 2.3. This is not an easy task: at this stage we can compare different solutions, and show the Lipschitz continuity on the initial data, only within a narrow class of functions with the same wave-front structure. Roughly speaking, by "wave-front structure" or "configuration" we refer here to the number of shocks of strength $> \varepsilon$, and to the order in which they interact. As shown in Step 1, we can construct suitable approximations by choosing artificial wave speeds λ_i^* and inserting the buffer zones (2.18) around each big shock. The distance between two approximations can then be estimated, but only if these approximations are obtained by the same choice of the λ_i^*, and by the insertion of the same number of shock layers. Of course, this cannot be the case for solutions with different wave-front configurations (Fig. 2.4).

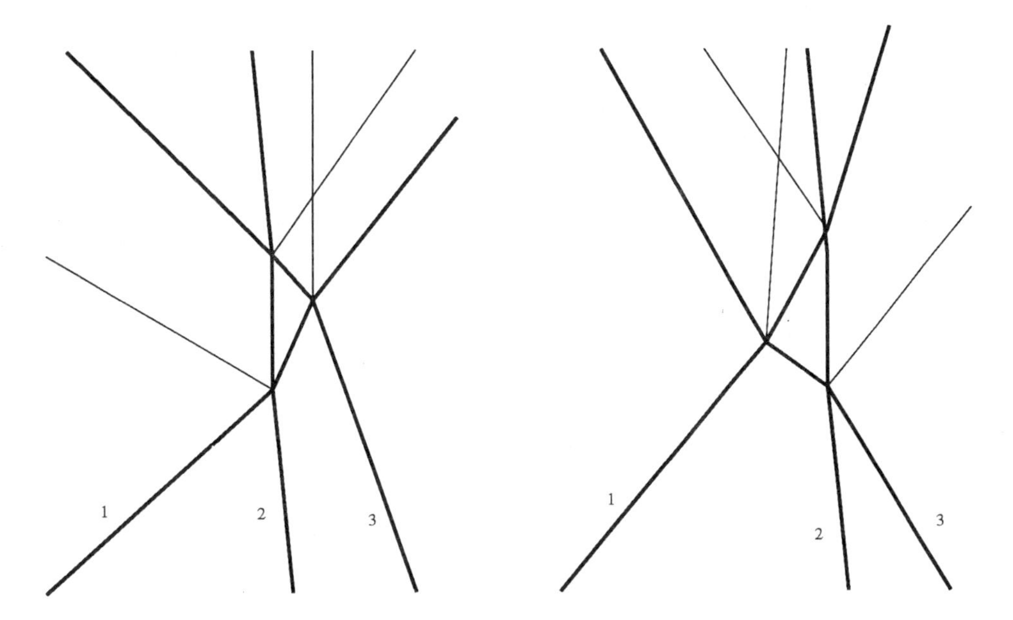

FIGURE 2.4

Our basic approach is the following. Consider a regular path of initial data $\gamma_0 : \theta \mapsto \bar{u}^\theta$, $\theta \in [0,1]$. For each $t > 0$, define the corresponding path

$$\gamma_t : \theta \mapsto S_t^\varepsilon \bar{u}^\theta. \tag{2.22}$$

First, we prove that the weighted length of γ does not increase in time if all solutions u^θ have the same wave-front structure. By the continuity of the semigroup S^ε, proved in Step 2, the result can be extended to the case where the configuration changes only at finitely many values of the parameter θ. Finally, we show that any path γ of solutions can be uniformly approximated by a path $\tilde{\gamma}$ such that the wave-front structure of the function $\tilde{u}^\theta = \tilde{\gamma}(\theta)$ changes only at finitely many parameter values.

To implement the above program, we introduce a set of conditions providing the structural stability of an ε-solution u. They will imply that any ε-solution u' sufficiently close to u has the same number of large shocks as u, interacting in the same order. In particular, u' can be obtained as limit of the same type of piecewise Lipschitz approximations used in the construction of u.

As a preliminary, given an ε-solution u and a point (t, x) with $t > 0$, define

$$u^\eta(t', x') \doteq u(t + \eta t',\ x + \eta x'), \qquad \omega_{(t,x)}(t', x') \doteq \lim_{\eta \to 0+} u^\eta(t', x'). \tag{2.23}$$

By the analysis in [**DP1, B-LF2**], the self-similar limit in (2.23) is well defined in $\mathbf{L}^1_{loc}$. On the upper half plane where $t' > 0$, $\omega_{(t,x)}$ coincides with the solution of the standard Riemann problem with data $u(t, x-)$, $u(t, x+)$. On the lower half plane where $t' < 0$, the function $\omega_{(t,x)}$ may contain a set of incoming waves, including shocks and centered compression waves of various families. We shall distinguish three cases.

CASE 1: The strength of all incoming waves is $< \varepsilon^2$.

CASE 2: There exists an incoming shock of strength $> \varepsilon^3$ while the total strength of all other incoming waves is $< \varepsilon^8$.

CASE 3: There exist two incoming shocks, both of strength $> \varepsilon^9$, while the total strength of all other incoming waves is $< \varepsilon^{20}$.

Observe that the above cases are not mutually exclusive, nor do they cover all possibilities. A suitable definition of structural stability can now be introduced.

DEFINITION 2.6. An ε-solution u of (1.1) is *Structurally Stable at the point* (t, x) if the corresponding function $\omega_{(t,x)}$ in (2.23) satisfies one of the Cases 1, 2, 3. We say that u is *Structurally Stable on* $[t^*, T]$ iff u is stable at each point (t, x) with $t^* \leq t \leq T$, $x \in \mathbb{R}$.

There are two typical examples where structural stability fails:

- Three shocks, each with strength $> \varepsilon$, interact at a single point (t, x).
- The point (t, x) is the center of a compression wave, of strength $> \varepsilon$.

Observe that, in both cases, an arbitrarily small perturbation may destroy the wave-front configuration of the solution u. The next proposition states that this does not happen for a structurally stable solution $\bar{u}$.

In the following, by $V\big(u(t);\ J\big)$ we denote the total strength of waves in $u(t)$ inside the set J.

PROPOSITION 2.7. *Let $\tilde{u}$ be an ε-solution of (1.1), structurally stable at the point $(\tau, \bar{x})$. Then there exists $r^* > 0$ such that, for every fixed $r \in]0, r^*]$, on the interval $J \doteq [\bar{x} - 7r^* - r,\ \bar{x} + 7r^* + r]$ the following holds. For every ε-solution u sufficiently close to $\tilde{u}$ in the $\mathbf{L}^1$ norm, one has:*

(i) *In Case 1, $u(\tau - r)$ satisfies*

$$V\big(u(\tau - r);\ J\big) \leq C\varepsilon^2. \tag{2.24}$$

(ii) *In Case 2, $u(\tau - r)$ has a shock of strength $> \varepsilon^3/2$, located at some point $y \in J$. Moreover,*

$$V\big(u(\tau - r);\ J \setminus \{y\}\big) \leq C\varepsilon^8. \tag{2.25}$$

(iii) *In Case 3, $u(\tau - r)$ has two approaching shocks, both of strength $> \varepsilon^9/2$, located at points $y, y' \in J$. Moreover,*

$$V\big(u(\tau - r);\ J \setminus \{y, y'\}\big) \leq C\varepsilon^{20}. \tag{2.26}$$

As usual, by C we denote a constant depending only on the system, and not on $u, \tilde{u}, r, r^*, \varepsilon \ldots$ On the other hand, it is understood that the conclusion of Proposition 2.7 holds for every solution u satisfying

$$\big\|u(t) - \tilde{u}(t)\big\|_{\mathbf{L}^1} < \varepsilon_* \qquad \forall t,$$

with $\varepsilon_* > 0$ possibly depending on r. For exact solutions of the system (1.1), or approximations obtained by wave-front tracking, a more general result in this direction is established in [**B-LF2**], relying on the decay estimates (1.21). All proofs remain valid for ε-solutions, without any change. In Section 7 we will show that the same conclusions (i)–(iii) also hold if u is a sufficiently accurate piecewise Lipschitz approximation, constructed according to our algorithm.

In the above setting, a trapezoid of the form

$$\Gamma \doteq \Big\{(t, x); \qquad t \in [\tau - r',\ \tau + r''],\ |x - \bar{x}| \leq 7r^* - (t - \tau)\Big\} \tag{2.27}$$

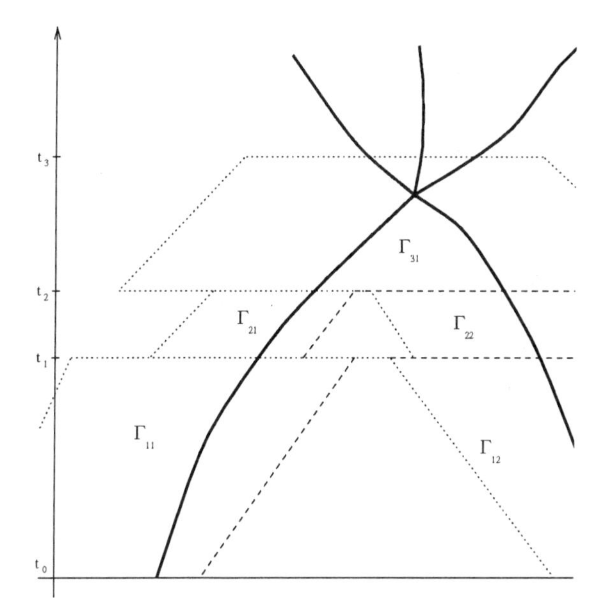

FIGURE 2.5

with $r', r'' \in]0, r^*]$ will be called a *Stabilizing Block* for the solution $\tilde{u}$ around the point $(\tau, \bar{x})$. Observe that, by Proposition 2.7, every solution u suitably close to $\tilde{u}$ in the $\mathbf{L}^1$ norm has the same wave-front configuration as $\tilde{u}$ inside the trapezoid Γ. This motivates our definition. We say that the stabilizing block Γ is of type 1, 2 or 3 according to the three cases considered in Proposition 2.7.

A straightforward compactness argument yields the following covering lemma.

LEMMA 2.8. *Let $M, t_* > 0$ be given and let $\tilde{u}$ be structurally stable on $[t_*, T]$. Then there exist times $0 < t_0 < \tau_1 < t_1 < \tau_2 < \cdots < \tau_N < t_N = T$, points x_{ij} and values $r_{ij} > 0$, $j = 1, \ldots, N_i$ with the following properties. For each $i = 1, \ldots, N$, the intervals $[x_{ij} - r_{ij},\ x_{ij} + r_{ij}]$ cover $[-M, M]$. Moreover,*

$$(2.28) \qquad [t_*, T] \subset [t_0, t_N], \qquad \max\{t_i - \tau_i,\ \tau_i - t_{i-1}\} \le \min_j r_{ij} \quad i = 1, \ldots N.$$

Each trapezoid

$$(2.29) \qquad \Gamma_{ij} \doteq \Big\{(t, x); \qquad t \in [t_{i-1},\ t_i],\ |x - x_{ij}| \le 7r_{ij} - (t - \tau_i)\Big\}$$

is a stabilizing block for $\tilde{u}$ around the point (τ_i, x_{ij}).

EXAMPLE 2.9. Consider an ε-solution $u = u(t, x)$ with the following wave-front structure (Fig. 2.5). Initially, u contains only one large shock, say at a point y_1, plus other small waves. As time progresses, a second shock is formed, say located at y_2. The strength of this shock increases continuously from 0 to some value $> \varepsilon$. At some time τ_2, the two shocks interact, generating three outgoing shocks.
In this case, u is structurally stable. A covering in terms of stabilizing blocks is illustrated in Fig. 2.5. The block Γ_{12} is of type 1, $\Gamma_{11}, \Gamma_{21}, \Gamma_{22}$ are of type 2, while Γ_{31} is of type 3.

We can now take the main step toward establishing the Lipschitz continuity of the semigroup S^ε. More precisely, we will show that S^ε is contractive w.r.t. a weighted distance defined as follows. Let u be a piecewise Lipschitz function having jumps at the points $x_1 < \cdots < x_N$. Assume that the ε-solution of the Riemann problem determined by the jump at x_α consists of a single shock in the k_α-th characteristic family, of strength $|\sigma_\alpha|$. For any $v \in \mathbf{L}^1$, define the components

$$v_i(x) \doteq \langle l_i(u(x)),\, v(x)\rangle, \qquad u_x^i(x) \doteq \langle l_i(u(x)),\, u_x(x)\rangle \qquad i = 1,\dots,n. \tag{2.30}$$

The weighted norm of a generalized tangent vector $(v,\xi) \in T_u = \mathbf{L}^1 \times \mathbb{R}^N$ is then defined as

$$\big\|(v,\xi)\big\|_u \doteq \sum_{i=1}^n \int_{-\infty}^{\infty} |v_i(x)| W_i^u(x)\, dx + \sum_{\alpha=1}^N |\xi_\alpha||\sigma_\alpha| W_{k_\alpha}^u(x_\alpha). \tag{2.31}$$

Here $W_i^u(x)$ is the weight assigned to an i-wave located at x. It has the form

$$W_i^u(x) \doteq 1 + \kappa_1 R_i^u(x) + \kappa_1\kappa_2 Q(u), \tag{2.32}$$

where

$$R_i^u(x) \doteq \left[\sum_{j\le i}\int_x^\infty + \sum_{j\ge i}\int_{-\infty}^x\right] |u_x^j(y)|dy + \left[\sum_{\substack{k_\alpha\le i\\ x_\alpha > x}} + \sum_{\substack{k_\alpha \ge i\\ x_\alpha < x}}\right] |\sigma_\alpha|, \tag{2.33}$$

$$\begin{aligned} Q(u) \doteq &\sum_{i\le j}\iint_{x<y} \big|u_x^j(x)\big|\big|u_x^i(y)\big|\, dxdy + \sum_{k_\alpha\le k_\beta,\ x_\alpha > x_\beta} |\sigma_\alpha\sigma_\beta| \\ &+ \sum_\alpha \left[\sum_{i\le k_\alpha} |\sigma_\alpha| \int_{x_\alpha}^\infty \big|u_x^i(x)\big|dx + \sum_{i\ge k_\alpha} |\sigma_\alpha| \int_{-\infty}^{x_\alpha} \big|u_x^i(x)\big|dx\right], \end{aligned} \tag{2.34}$$

and κ_1, κ_2 are suitably large constants. Thinking of $\big|u_x^i(x)\big|dx$ as the strength of an infinitesimal i-wave in u located at x, the weight (2.32) can thus be interpreted as

$$\begin{aligned} W_i^u(x) = 1 &+ \kappa_1\big[\text{amount of waves in } u \text{ which approach an } i\text{-wave located at } x\big] \\ &+ \kappa_1\kappa_2\big[\text{global interaction potential of waves in } u\big]. \end{aligned}$$

A general formula for the weighted length of a tangent vector, valid for functions u with arbitrary jumps, will be given in Section 8. Here and in the sequel there will be slight variations in the definitions of Q, due to the fact that couples of waves of the same family may or may not be regarded as approaching. The different choices made at various stages of the paper aim at simplifying the computations in the proofs. They are all essentially equivalent in that the basic interaction estimates (1.9) always hold.

In turn, the weighted length of a piecewise regular path can then be defined as in (1.47), by integrating the norm of a tangent vector. To make sense of the weighted length of an arbitrary path $\gamma\colon \theta \mapsto u^\theta$ in the case where the u^θ are general BV functions, we set

$$\|\gamma\|_\star \doteq \lim_{\varepsilon\to 0+} \inf\Big\{\|\gamma'\|_\star;\ \gamma' \text{ piec. regular path, } \big\|\gamma'(\theta) - \gamma(\theta)\big\|_{\mathbf{L}^1} < \varepsilon\ \forall\theta\Big\}. \tag{2.35}$$

PROPOSITION 2.10. *Let* $\gamma_0 \colon \theta \mapsto \bar{u}^\theta$ *be a regular path of initial data, say defined for* $\theta \in \Theta \doteq [a,b]$. *For some value* $\bar{\theta}$, *assume that the corresponding* ε*-solution* $u^{\bar{\theta}}$ *of (1.1) is structurally stable on* $[0,T]$. *Then there exists* $\rho > 0$ *such that the weighted length of the path* $\theta \mapsto S_t^\varepsilon \bar{u}^\theta$, *restricted to* $\theta \in [\theta', \theta'']$, *is a non-increasing function of time, for every subinterval* $[\theta', \theta''] \subseteq [\bar{\theta} - \rho,\ \bar{\theta} + \rho]$.

As a first step toward the proof of Proposition 2.10, we show that the conclusion holds in the special case where the solutions u^θ vary with θ only within a single isolating block.

LEMMA 2.11. *Let* Γ *be as in (2.27) and let* $\theta \mapsto u^\theta$ *be a family of* ε*-solutions of (1.1), defined for* $t \in [t', t''] \doteq [\tau - r',\ \tau + r'']$, $\theta \in \Theta \doteq [a,b]$. *Assume that, at the initial time* $t' = \tau - r'$, *all functions* $u^\theta(t', \cdot)$ *satisfy one of the conclusions of Proposition 2.7: either (i), or (ii), or (iii). Moreover, assume that for all* θ *one has*

$$u^\theta(t', x) = u^{\bar{\theta}}(t', x) \qquad x \notin [\bar{x} - r^*,\ \bar{x} + r^*]. \tag{2.36}$$

Then the weighted length of the path $\gamma_{t''} \colon \theta \mapsto u^\theta(t'', \cdot)$ *is smaller or equal to the length of the path* $\gamma_{t'} \colon \theta \mapsto u^\theta(t', \cdot)$.

The proof of Lemma 2.11 is achieved by constructing a path γ' of piecewise Lipschitz approximate solutions, arbitrarily close to γ, whose length does not increase in time. Observe that, inside the trapezoid Γ, all solutions u^θ have the same wave-front structure as $u^{\bar{\theta}}$. Namely:

- in Case 1 no large shock is present,
- in Case 2 there is exactly one large shock,
- in Case 3 there are initially two large incoming shocks. At some time t these shocks interact, generating a number of outgoing waves determined by the corresponding Riemann problem.

Inside the region Γ, the construction of approximate solutions, continuously depending on the initial data, can thus be carried out by the same algorithm used in the proof of Proposition 2.4. Actually, in Cases 1 and 2, we are dealing with a set of functions all contained in one of the domains $\mathcal{D}_{(u^-, u^+)}$ considered at (2.8) (if suitably extended outside Γ). Case 3, on the other hand, forces us to consider a domain $\mathcal{D}$ of functions which possibly contain two large approaching shocks. In this more general case, the construction of piecewise Lipschitz approximate solutions will be given in Section 8.

Outside the stabilizing block Γ, our approximate solutions will be constructed by wave-front tracking. Observe that, in this outer region, nothing is known about the wave-front structure of the functions u^θ. However, by (2.36) and finite propagation speed, all these solutions coincide. Their continuous dependence on the parameter θ is thus trivial.

In order to apply Lemma 2.11, we shall need to replace an arbitrary path γ with a new path $\tilde{\gamma} \colon \theta \mapsto \tilde{u}^\theta$ such that the values $\tilde{u}^\theta(x)$ locally vary with θ only inside one single stabilizing block. A suitable localization procedure is described below.

DEFINITION 2.12. A path $\gamma \colon \theta \mapsto u^\theta$ has *Localized Variation* if, for every $\varepsilon^* > 0$ and every θ^*, there exists $\delta > 0$ and a point x^* such that

$$u^\theta(x) = u^{\theta^*}(x) \qquad \text{whenever } |\theta - \theta^*| < \delta,\ |x - x^*| \geq \varepsilon^*. \tag{2.37}$$

In other words, as θ varies in a neighborhood of θ^*, the values of u^θ are allowed to change only inside a small neighborhood of some point x^*. For example, if u, v are two distinct continuous functions, then the path $\theta \mapsto u \cdot \chi_{]-\infty,\theta]} + v \cdot \chi_{]\theta,\infty[}$ has localized variation, while the path $\theta \mapsto \theta u + (1-\theta)v$ does not. The approximation of piecewise regular paths with paths having localized variation will play a key role in the sequel.

LEMMA 2.13. *Let $\gamma: \theta \mapsto u^\theta$ be a piecewise regular path. Then, for every $\varepsilon^* > 0$, there exists another piecewise regular path γ' with localized variation such that*

$$\|\gamma'\|_\star \leq \|\gamma\|_\star + \varepsilon^*, \qquad \big\|\gamma'(\theta) - \gamma(\theta)\big\|_{\mathbf{L}^1} < \varepsilon^* \qquad \forall \theta. \tag{2.38}$$

The proof of Proposition 2.10 now goes as follows. For t in a small interval $[0, t_0]$, all solutions u^θ remain piecewise Lipschitz, and the result is an immediate consequence of Proposition 1.7. Choose M large enough so that all supports of the functions $u^\theta(t,\cdot)$ are contained inside $[-M, M]$. We then apply Lemma 2.8 to the structurally stable solution $u^{\bar\theta}$ and obtain a covering of the rectangle $[t_0, T] \times [-M, M]$ in terms of stabilizing blocks Γ_{ij}. Let (τ_i, x_{ij}) and r_{ij} be as in (2.29). For some $\rho > 0$ small and all $\theta \in \Theta \doteq [\bar\theta - \rho,\ \bar\theta + \rho]$, we can thus assume that every suitably accurate approximation to a solution u^θ has the same wave-front structure as $u^{\bar\theta}$, on each trapezoid Γ_{ij}.

By Lemma 2.13, the path $\gamma_{t_0}: \theta \mapsto S^\varepsilon_{t_0}\bar u^\theta$ can be replaced by a new path $\gamma^+_{t_0}: \Theta \mapsto BV$, having localized variation and almost the same length as γ_{t_0}. By compactness, we can cover the interval Θ with finitely many subintervals $\Theta_k \doteq [\theta_{k-1}, \theta_k]$, $k = 1, \ldots, m_1$, so that, as θ varies inside Θ_k, the values of $u^\theta(t_0+, \cdot) \doteq \gamma^+_{t_0}(\theta)$ vary inside a single interval, say $I_{1j} \doteq [x_{1j} - r_{1j},\ x_{1j} + r_{1j}]$. Here $j = j(k)$.

By applying Lemma 2.11 to each subinterval Θ_k, we conclude that the weighted length of the path

$$\gamma^-_{t_1}: \theta \mapsto S^\varepsilon_{t_1 - t_0}\big(\gamma^+_{t_0}(\theta)\big)$$

is less than or equal to the weighted length of $\gamma_{t_0^-}$.

We now apply Lemma 2.13, replacing $\gamma^-_{t_1}$ with a new path $\gamma^+_{t_1}$, having localized variation and almost the same length as $\gamma_{t_1^-}$. Then we use Lemma 2.11 to estimate the length of the path $\gamma^-_{t_2} \doteq S^\varepsilon_{t_2 - t_1}(\gamma^+_{t_1})$, etc... Since at every step the increase in the weighted length $\|\gamma^+_{t_i}\|_\star - \|\gamma^-_{t_i}\|_\star$ can be kept arbitrarily small, after N steps we obtain a path $\gamma^-_{t_N}$, arbitrarily close to $\gamma_T: \theta \mapsto S^\varepsilon_T \bar u^\theta$, and whose weighted length is bounded by the length of γ_0 up to an arbitrary small correction. This achieves the proof.

From Proposition 2.10 and the continuity of the semigroup S^ε, one obtains

COROLLARY 2.14. *If the ε-solution u^θ is structurally stable on $]0, T]$ for all but finitely many values of θ, then the weighted length of the path $\gamma_t: \theta \mapsto u^\theta$ is non-increasing in time.*

The proof of Theorem 2.3 is completed by showing that the above assumption of structural stability can be removed, relying on a perturbation argument.

PROPOSITION 2.15. *Let $\gamma_0: \theta \mapsto \bar u^\theta$ be a regular path of initial conditions. Then the conclusion of Proposition 2.10 remains valid, even without assuming the structural stability of the ε-solution $u^{\bar\theta}$.*

The proof is based on the following argument. Call Θ^* the set of all values of θ for which u^θ is structurally unstable. If $\bar\theta \notin \Theta^*$, or if $\bar\theta$ is an isolated point of Θ^*, the result is clear. Consider the remaining case, where $\bar\theta$ is a limit point of Θ^*. Observe that there exists only finitely many points (t_ℓ, x_ℓ) in the t-x plane where $u^{\bar\theta}$ is unstable. Indeed, by Definition 2.6, at every such point an amount of interaction $> \varepsilon^{18}$ must take place.

To fix the ideas, assume, for example, that for infinitely many $\theta \in \Theta^*$ in a neighborhood of $\bar\theta$, the corresponding ε-solution u^θ contains three large shocks interacting at the single point (t^θ, x^θ), with $(t^\theta, x^\theta) \to (t^{\bar\theta}, x^{\bar\theta})$ as $\theta \to \bar\theta$. At a suitable time $\tau < t^{\bar\theta}$ we can then perform an arbitrarily small perturbation of the path $\gamma_\tau : \theta \mapsto u^\theta(\tau)$, in such a way that all but finitely many of the perturbed solutions are structurally stable. More precisely, given any $\varepsilon' > 0$, we consider a smooth scalar map $\varphi = \varphi(\theta, x)$ with

$$\|\varphi\|_{C^3} < \varepsilon' \tag{2.39}$$

and construct a new path $\tilde\gamma_\tau : \theta \mapsto \tilde u^\theta(\tau)$, where

$$\tilde u^\theta(\tau, x) = u^\theta\big(\tau-,\ x + \varphi(\theta, x)\big). \tag{2.40}$$

Because of (2.39), the new path will be close to the old one and have almost the same weighted length. In addition, by (2.40) the locations of the three large shocks in each u^θ will be slightly shifted. Therefore, an application of the coarea formula **[E-G]** will show that, for a "generic" function φ, these three shocks will no longer interact at a single point, but only two at a time, for all except finitely many values of the parameter θ. The previous estimates can thus be applied to the perturbed path $\tilde\gamma$, for $t > \tau$.

This will complete the proof of Theorem 2.3. Letting $\varepsilon \to 0$, since the constants $L, \eta_0 > 0$ are independent of ε, we establish Theorem 1.1.

CHAPTER 3

Construction of local semigroups

Throughout the following we fix $\varepsilon > 0$ and consider a fixed set of right and left eigenvectors $r_i(u), l_i(u)$ of $A(u) = DF(u)$, normalized as in (1.5)-(1.6). Let two states u^-, u^+ be given, and call $\omega = \omega(t,x)$ the ε-approximate solution of the Riemann problem with initial data (2.6). Aim of this section is to construct a semigroup S whose trajectories are ε-approximate solutions of (1.1) according to Definition 1.9, and whose domain contains all sufficiently small perturbations of the Riemann data (2.6).

Let $\omega_0 = u^-, \omega_1, \ldots, \omega_n = u^+$ be the intermediate states in the self-similar ε-solution ω. Let $\mathcal{S} = \{j_1, \ldots, j_\nu\} \subseteq \{1, \ldots, n\}$ be the set of indices i for which the wave connecting ω_{i-1} with ω_i is a shock of strength $|\sigma_i| \geq 2\varepsilon$. By choosing $\eta < \varepsilon^2$ sufficiently small in (2.8), we can assume that every function $u \in \mathcal{D}$ contains ν shocks with strength $> \varepsilon$, one for each of the families $j_1, \ldots, j_\nu$, located at points $y_{j_1} < \cdots < y_{j_\nu}$, plus possibly other shocks (of different families) with strength $< 3\varepsilon$. Define the characteristic speeds

$$
\begin{aligned}
&\lambda_i^* \doteq \lambda_i(\omega_{i-1}), \qquad \lambda_i^{**} \doteq \lambda_i(\omega_i) \qquad &&\text{if } i \in \mathcal{S}, \\
&\lambda_i^* \doteq \lambda_i^{**} \doteq \lambda_i(u^-) &&\text{if } i \notin \mathcal{S}.
\end{aligned} \tag{3.1}
$$

Given an interval $[0,T]$ and $\delta_1 > 0$, we will construct an approximate solution $u = u(t,x)$ with the following properties. There exists a finite partition of $[0,T]$ into subintervals $J_\ell = [\tau_\ell,\ \tau_{\ell+1}[$, such that the restriction of u to each strip $J_\ell \times \mathbb{R}$ is piecewise Lipschitz continuous, with jumps located along finitely many lines $x = y_\alpha(t)$. At each time τ_ℓ, a restarting procedure is used, producing a new function $u(\tau_\ell, \cdot)$, suitably close to $u(\tau_\ell -, \cdot)$ in a sense which will be made precise later.

Each subinterval J_ℓ is entirely contained in some interval of the form

$$
I_{m,h} = \left[\Big(m + \frac{h-1}{n}\Big)\delta_1,\ \Big(m + \frac{h}{n}\Big)\delta_1\right] \qquad (m \geq 0,\ h \in \{1, \ldots, n\}). \tag{3.2}
$$

The restriction of u to the strip $J_\ell \times \mathbb{R}$ contains ν large jumps of strength $|\sigma| > \varepsilon$, corresponding to the large shocks in the self-similar solution ω of the Riemann problem. These are located at the points $y_{j_1}(t) < \cdots < y_{j_\nu}(t)$ and occur respectively in the $j_1, \ldots, j_\nu$-characteristic families. The corresponding ε-Rankine-Hugoniot conditions (2.3)–(2.5) hold at each one of these points.

In addition, if $J_\ell \subseteq I_{m,h}$, the function u has finitely many small h-shocks, say located along the lines $x = y_\beta(t)$, $\beta \in \mathcal{S}'$ (here $\mathcal{S}'$ is some index set, disjoint from $\mathcal{S}$). These small shocks satisfy

$$
u(y_\beta +) = R_h(\sigma_\beta)\big(u(y_\beta -)\big) \qquad \beta \in \mathcal{S}' \tag{3.3}
$$

for some $\sigma_\beta \in [-3\varepsilon, 0[$, so that the left and right states at y_β lie on the same h-rarefaction curve. To describe the evolution equation satisfied by u outside the shocks, we introduce the matrix $A^{(h)} = A^{(h)}(t,x,u)$, whose eigenvectors coincide with the eigenvectors r_i, l_i of $A(u)$, and whose eigenvalues $\lambda_i^{(h)} = \lambda_i^{(h)}(t,x,u)$ are defined as follows:

$$\lambda_i^{(h)} = \begin{cases} \lambda_i^* & \text{if } i \notin \mathcal{S}, \\ \lambda_i^* & \text{if } i \in \mathcal{S},\ x < y_i(t), \\ \lambda_i^{**} & \text{if } i \in \mathcal{S},\ x > y_i(t), \end{cases} \tag{3.4}$$

for all $i \neq h$, so that $n-1$ characteristic fields are linearly degenerate. The definition of the h-th eigenvalue requires more care. Indeed, we want the h-field to be genuinely nonlinear, except inside some artificial "shock layers" of width $\delta_2 > 0$ around the large shocks, where $\lambda_h^{(h)}$ will be constant. For $\alpha \in \mathcal{S}$, consider the lines

$$y_\alpha^*(t) = y_\alpha(t) - \delta_2, \qquad y_\alpha^{**}(t) = y_\alpha(t) + \delta_2. \tag{3.5}$$

If $h \in \mathcal{S}$, we then define

$$\lambda_h^{(h)} = \begin{cases} \lambda_h^* + n\big(\lambda_h(\omega_{\alpha-1}) - \lambda_h^*\big) & \text{if } x \in \big[y_\alpha^*(t),\ y_\alpha(t)\big[\text{ for some } \alpha \in \mathcal{S}, \\ \lambda_h^{**} + n\big(\lambda_h(\omega_\alpha) - \lambda_h^{**}\big) & \text{if } x \in \big]y_\alpha(t),\ y_\alpha^{**}(t)\big] \text{ for some } \alpha \in \mathcal{S}, \\ \lambda_h^* + n\big(\lambda_h(u) - \lambda_h^*\big) & \text{if } x < y_h(t) \text{ and } x \notin \bigcup \big[y_\alpha^*(t),\ y_\alpha^{**}(t)\big], \\ \lambda_h^{**} + n\big(\lambda_h(u) - \lambda_h^{**}\big) & \text{if } x > y_h(t) \text{ and } x \notin \bigcup \big[y_\alpha^*(t),\ y_\alpha^{**}(t)\big]. \end{cases} \tag{3.6}$$

In the case $h \notin \mathcal{S}$, the last two cases in (3.6) are replaced by

$$\lambda_h^{(h)} = \lambda_h^* + n\big(\lambda_h(u) - \lambda_h^*\big) \qquad \text{if} \quad x \notin \bigcup \big[y_\alpha^*(t),\ y_\alpha^{**}(t)\big]. \tag{3.7}$$

The definitions (3.4)–(3.7) completely determine the matrix $A^{(h)}$. If $J_\ell \subseteq I_{m,h}$, then on the strip $J_\ell \times \mathbb{R}$ we require that the piecewise Lipschitz function u be a solution of the quasilinear hyperbolic system

$$u_t + A^{(h)}(t,x,u)u_x = 0 \tag{3.8}$$

outside the shock lines. Finally, recalling (3.3), we assign the speed of a small h-shock located at y_β:

$$\dot y_\beta(t) = \frac{1}{|\sigma_\beta|} \int_{\sigma_\beta}^0 \lambda_h^{(h)}\Big(t,\ y_\beta,\ R_h(s)\big(u(y_\beta -)\big)\Big)\, ds. \tag{3.9}$$

The quasilinear system (3.8), together with the ε-Rankine-Hugoniot equations (2.3)-(2.4) valid for the big shocks at y_α, $\alpha \in \mathcal{S}$, and with the relations (3.3), (3.9) valid for the small h-shocks at y_β, $\beta \in \mathcal{S}'$, entirely determine the evolution of our piecewise Lipschitz approximate solution u, within each time interval $J_\ell = [\tau_\ell,\ \tau_{\ell+1}[$. The piecewise Lipschitz regularity of u will be preserved until one of the following situations occurs:

- A gradient catastrophe takes place, in the h-family.
- A small h-shock interacts with one of the large shocks.

Before this happens, a restarting procedure will be used, replacing $u(\tau_{\ell+1}-)$ with a new (better behaved) function. An additional restarting is performed at the end of each interval $I_{m,h}$ in (3.2), where the evolution of u changes type. All these restarting procedures will be described in Section 4. In the present section we study solutions of the above evolution equations on a time interval J_ℓ bounded by two consecutive restarting times. A-priori bounds will be obtained on the total strength of waves and on the weighted norm of generalized tangent vectors.

We begin by deriving a set of evolution equations for the gradient components $u_x^i \doteq \langle l_i(u),\ u_x\rangle$ and for the strength of the shocks, valid on the time intervals $I_{m,h}$ of the form (3.2). Using a continuous version of the Glimm interaction functional, this will imply a uniform bound on the total variation of $u(t,\cdot)$.

As in [**B1, B-M1**], from (3.8) we obtain

$$
\begin{aligned}
(u_x^i)_t + \left(\lambda_i^{(h)} u_x^i\right)_x &= \sum_{j<k} \left(\lambda_k^{(h)} - \lambda_j^{(h)}\right) \langle l_i,\ [r_k, r_j]\rangle u_x^j u_x^k \\
&\doteq \sum_{j<k} G_{ijk}(u) u_x^j u_x^k,
\end{aligned}
\tag{3.10}
$$

valid outside the shock lines y_α, $\alpha \in \mathcal{S} \cup \mathcal{S}'$ and outside the lines y_α^*, y_α^{**}, $\alpha \in \mathcal{S}$, where the h-characteristic speed $\lambda_h^{(h)}$ is discontinuous. As usual, $[r_k, r_j] \doteq \nabla r_j \cdot r_k - \nabla r_k \cdot r_j$ denotes the Lie bracket of the vector fields r_k, r_j.

Let k_α be the family of the shock at y_α. According to our previous notation, we thus have $k_\alpha = \alpha$ if $\alpha \in \mathcal{S}$, $k_\alpha = h$ if $\alpha \in \mathcal{S}'$. Define the sets $\mathcal{I}$ and $\mathcal{O}$ (incoming and outgoing) of signed indices

$$
\begin{aligned}
\mathcal{I} &= \{i^+;\ i \leq k_\alpha\} \cup \{i^-;\ i \geq k_\alpha\}, \\
\mathcal{O} &= \{j^-;\ j < k_\alpha\} \cup \{j^+;\ j > k_\alpha\}.
\end{aligned}
\tag{3.11}
$$

In a neighborhood of $\big(u(y_\alpha-),\ u(y_\alpha+)\big)$ we use the coordinate system

$$
u^- = u(y_\alpha-) + \sum_{i=1}^n r_i\big(u(y_\alpha-)\big) w_i^-, \qquad u^+ = u(y_\alpha+) + \sum_{i=1}^n r_i\big(u(y_\alpha+)\big) w_i^+.
$$

Recalling (2.2), the relation (2.3) can now be written as a system of $n-1$ scalar equations in the $2n$ variables w_i^-, w_i^+, $i = 1, \ldots, n$. By the implicit function theorem, the equations (2.3) can then be solved for the $n-1$ outgoing components:

$$
w_j^\pm = W^j(w^{\mathcal{I}}) \qquad j^\pm \in \mathcal{O}
\tag{3.12}
$$

where $w^{\mathcal{I}}$ denotes the set of incoming components.

In order not to interrupt the flow of the main argument, in the following we state without proof a number of a priori estimates. Most of these estimates are entirely standard. The others will be proved in the Appendix. For notational convenience, we denote by C a constant which depends only on the system (1.1), and not on the parameters $\varepsilon, \delta_1, \delta_2$ or on the particular solution. In a chain of inequalities, the value of C may change from one term to the next. Constants which play a distinguished role will be written as $C_0, C_1, \ldots$

Let $\sigma_\alpha < 0$ be the size of the shock at y_α. The derivatives of the functions W^j at $w^{\mathcal{I}} = 0$ satisfy

$$
\left|\frac{\partial W^j}{\partial w_i^\pm}\right| \leq C \cdot |\sigma_\alpha| \qquad i \neq j,\ i^\pm \in \mathcal{I},\ j^\pm \in \mathcal{O},
\tag{3.13}
$$

$$
\left|\frac{\partial W^i}{\partial w_i^\pm} - 1\right| \leq C \cdot |\sigma_\alpha| \qquad i \neq k_\alpha,\ i^\pm \in \mathcal{I},\ i^\mp \in \mathcal{O},
\tag{3.14}
$$

$$
\left|\frac{\partial W^j}{\partial w_{k_\alpha}^\pm}\right| \leq C \cdot |\sigma_\alpha|^2 \qquad j^\pm \in \mathcal{O}.
\tag{3.15}
$$

Concerning the size and speed of the shock at y_α, we have the estimates

$$\left|\frac{\partial\sigma_\alpha}{\partial w_i^\pm}\right| \le C\cdot|\sigma_\alpha| \qquad i^\pm\in\mathcal{I},\ i\neq k_\alpha, \tag{3.16}$$

$$\left|\frac{\partial\sigma_\alpha}{\partial w_{k_\alpha}^\pm}\mp 1\right| \le C\cdot|\sigma_\alpha|^2, \tag{3.17}$$

$$\left|\frac{\partial\dot y_\alpha}{\partial w_i^\pm}\right| \le C \qquad i^\pm\in\mathcal{I},\ i\neq k_\alpha, \tag{3.18}$$

$$\left|\frac{\partial\dot y_\alpha}{\partial w_{k_\alpha}^-} - \frac{\lambda_{k_\alpha}^{(h)}(y_\alpha-)-\dot y_\alpha}{|\sigma_\alpha|}\right| \le C\cdot|\sigma_\alpha|, \tag{3.19}$$

$$\left|\frac{\partial\dot y_\alpha}{\partial w_{k_\alpha}^+} - \frac{\dot y_\alpha-\lambda_{k_\alpha}^{(h)}(y_\alpha+)}{|\sigma_\alpha|}\right| \le C\cdot|\sigma_\alpha|. \tag{3.20}$$

In the case of a small h-shock, from the relations (3.3)–(3.5) we deduce

$$\frac{\partial W^j}{\partial w_{k_\alpha}^\pm} = 0, \qquad \frac{\partial\sigma_\alpha}{\partial w_{k_\alpha}^\pm} = \pm 1, \tag{3.21}$$

$$\frac{\partial\dot y_\alpha}{\partial w_{k_\alpha}^-} = \frac{\lambda_{k_\alpha}^{(h)}(y_\alpha-)-\dot y_\alpha}{|\sigma_\alpha|}, \qquad \frac{\partial\dot y_\alpha}{\partial w_{k_\alpha}^+} = \frac{\dot y_\alpha-\lambda_{k_\alpha}^{(h)}(y_\alpha+)}{|\sigma_\alpha|}. \tag{3.22}$$

By $\lambda_i^{(h)}(y_\alpha+)$, $\lambda_i^{(h)}(y_\alpha-)$ we denote respectively the right and left limits of the function $\lambda_i^{(h)}(u(x))$ at $x=y_\alpha$.

Next, call u_x^{i-}, u_x^{i+} respectively the i-th component of u_x to the left and to the right of the shock. From the jump equations (2.3) we derive a family of $n-1$ linear relations

$$u_x^{j\pm} = U^j(u_x^{\mathcal{I}}). \tag{3.23}$$

We now observe that the gradient components $u_x^{i\pm}$ satisfy (3.23) iff the components

$$w_i^- \doteq \big(\lambda_i^{(h)}(y_\alpha-)-\dot y_\alpha\big)u_x^{i-}, \qquad w_i^+ \doteq \big(\dot y_\alpha-\lambda_i^{(h)}(y_\alpha+)\big)u_x^{i+}$$

satisfy the corresponding equations (3.12), linearized at $w^{\mathcal{I}}=0$. By strict hyperbolicity, from the estimates concerning the components $w_i^\pm$ it thus follows

$$\left|\frac{\partial U^j}{\partial u_x^{i\pm}}\right| \le C\cdot|\sigma_\alpha| \qquad i\neq j,\ i^\pm\in\mathcal{I},\ j^\pm\in\mathcal{O}, \tag{3.24}$$

$$\left|\frac{\partial U^i}{\partial u_x^{i\pm}} - 1\right| \le C\cdot|\sigma_\alpha| \qquad i\neq k_\alpha,\ i^\pm\in\mathcal{I},\ i^\mp\in\mathcal{O}, \tag{3.25}$$

$$\left|\frac{\partial U^j}{\partial u_x^{k_\alpha\pm}}\right| \le C\cdot|\sigma_\alpha|^2\big|\lambda_{k_\alpha}^{(h)}(y_\alpha\pm)-\dot y_\alpha\big| \qquad j^\pm\in\mathcal{O}. \tag{3.26}$$

In case of a small h-shock located at $x = y_\alpha(t)$, $\alpha \in \mathcal{S}'$, the estimates (3.24), (3.25) still hold, while (3.21) implies

$$\left|\frac{\partial U^{j\pm}}{\partial u_x^{h\pm}}\right| = 0 \qquad j^\pm \in \mathcal{O}. \tag{3.27}$$

Finally, if $\alpha \in \mathcal{S}$, across the two lines $x = y_\alpha^*(t)$, $x = y_\alpha^{**}(t)$ the function u is continuous while the only component of the gradient that suffers a jump is u_x^h. Observing that $\dot{y}_\alpha = \dot{y}_\alpha^* = \dot{y}_\alpha^{**}$, one checks that the left and right values of u_x^h are related by

$$\big(\lambda_h^{(h)}(y_\alpha^*+) - \dot{y}_\alpha\big)u_x^{h+} = \big(\lambda_h^{(h)}(y_\alpha^*-) - \dot{y}_\alpha\big)u_x^{h-} \qquad \text{at } x = y_\alpha^*, \tag{3.28}$$

and a similar equality holds at $x = y_h^{**}$.

Next, consider the time derivative of the size $\sigma_\alpha < 0$ of a shock at $x = y_\alpha(t)$. For $\alpha \in \mathcal{S}$ there holds

$$\begin{aligned} &\left|\dot{\sigma}_\alpha - \big(\lambda_{k_\alpha}^{(h)}(y_\alpha-) - \dot{y}_\alpha\big)u_x^{k_\alpha-} - \big(\dot{y}_\alpha - \lambda_{k_\alpha}^{(h)}(y_\alpha+)\big)u_x^{k_\alpha+}\right| \\ &\qquad \le C\cdot\left(|\sigma_\alpha|^3\big(|u_x^{k_\alpha+}| + |u_x^{k_\alpha-}|\big) + |\sigma_\alpha| \sum_{i^\pm\in\mathcal{I}, i\neq k_\alpha} |u_x^{i^\pm}|\right). \end{aligned} \tag{3.29}$$

Moreover, for $\alpha \in \mathcal{S}'$, the size $\sigma_\alpha \in [-3\varepsilon, 0]$ of a small h-shock located at $x = y_\alpha(t)$ satisfies the sharper estimate

(3.30)
$$\left|\dot{\sigma}_\alpha - \big(\lambda_h^{(h)}(y_\alpha-) - \dot{y}_\alpha\big)u_x^{h-} - \big(\dot{y}_\alpha - \lambda_h^{(h)}(y_\alpha+)\big)u_x^{h+}\right| \le C\cdot|\sigma_\alpha| \sum_{i^\pm\in\mathcal{I}, i\neq h} |u_x^{i\pm}|.$$

It might help the reader to compare (3.29) and (3.30) with the identity (1.43), valid in the scalar case.

To obtain a bound on the total variation, define the total strength of waves as

$$V(u) \doteq \sum_i \int_{-\infty}^{\infty} |u_x^i(x)|\, dx + \sum_{\alpha\in\mathcal{S}\cup\mathcal{S}'} |\sigma_\alpha|. \tag{3.31}$$

The interaction potential is defined as

$$\begin{aligned} Q(u) \doteq &\sum_{i<j} \iint_{x<x'} |u_x^j(x)||u_x^i(x')|\, dx dx' \\ &+ \sum_{\alpha\in\mathcal{S}\cup\mathcal{S}'} |\sigma_\alpha| \left[\sum_{i<k_\alpha} \int_{y_\alpha}^{\infty} |u_x^i(x)|\, dx + \sum_{i>k_\alpha} \int_{-\infty}^{y_\alpha} |u_x^i(x)|\, dx\right] \\ &+ \sum_{\substack{\alpha,\beta\in\mathcal{S}\cup\mathcal{S}' \\ k_\alpha>k_\beta,\, y_\alpha<y_\beta}} |\sigma_\alpha\sigma_\beta| \\ &+ \sum_{i\in\mathcal{S}} |\sigma_i| \left(\int_{-\infty}^{\infty} |u_x^i(x)|\, dx + \sum_{\beta\in\mathcal{S}',\, k_\beta=i} |\sigma_\beta|\right). \end{aligned} \tag{3.32}$$

Observe that in (3.32) two waves of the same family are never regarded as "approaching", except in the case where one of them is a large shock. Such a definition is natural in the present context. Indeed, due to the coincidence of shock and rarefaction curves (for small shock strengths), the interaction of small waves of the same family does not increase the total amount of waves.

The instantaneous amount of interaction is

$$\Lambda(u) \doteq \tilde{\Lambda}(u) + \sum_{\alpha \in \mathcal{S} \cup \mathcal{S}'} \Lambda_\alpha(u), \tag{3.33}$$

with

$$\tilde{\Lambda}(u) \doteq \sum_{i<j} \int_{-\infty}^{\infty} \left(\lambda_j^{(h)}(u(x)) - \lambda_i^{(h)}(u(x)) \right) \left| u_x^i(x) \right| \left| u_x^j(x) \right| \, dx, \tag{3.34}$$

and

$$\begin{aligned} \Lambda_\alpha(u) \doteq |\sigma_\alpha| \Big[&\sum_{i \geq k_\alpha} \big(\lambda_i^{(h)}(y_\alpha -) - \dot{y}_\alpha \big) \left| u_x^i(y_\alpha -) \right| \\ &+ \sum_{i \leq k_\alpha} \big(\dot{y}_\alpha - \lambda_i^{(h)}(y_\alpha +) \big) \left| u_x^i(y_\alpha +) \right| \Big], \\ \Lambda_\alpha(u) \doteq |\sigma_\alpha| \Big[&\sum_{i > k_\alpha} \big(\lambda_i^{(h)}(y_\alpha -) - \dot{y}_\alpha \big) \left| u_x^i(y_\alpha -) \right| \\ &+ \sum_{i < k_\alpha} \big(\dot{y}_\alpha - \lambda_i^{(h)}(y_\alpha +) \big) \left| u_x^i(y_\alpha +) \right| \Big], \end{aligned} \tag{3.35}$$

in the cases $\alpha \in \mathcal{S}$ and $\alpha \in \mathcal{S}'$, respectively. Using the bounds (3.24)–(3.30), a lengthy but straightforward computation yields

$$\begin{aligned} \frac{d}{dt} V(u) \leq &\sum_i \sum_{j<k} \left| G_{ijk}(u) \right| \left| u_x^j \right| \left| u_x^k \right| + \sum_{\alpha \in \mathcal{S} \cup \mathcal{S}'} |\dot{\sigma}_\alpha| \\ &+ \sum_{\alpha \in \mathcal{S} \cup \mathcal{S}'} \left(\sum_{j^\pm \in \mathcal{O}} \left| \lambda_j^{(h)}(y_\alpha \pm) - \dot{y}_\alpha \right| \left| u_x^{j\pm} \right| - \sum_{j^\pm \in \mathcal{I}} \left| \lambda_j^{(h)}(y_\alpha \pm) - \dot{y}_\alpha \right| \left| u_x^{i\pm} \right| \right) \\ \leq &C_4 \Lambda(u) \end{aligned}$$

for some constant C_4. Moreover,

$$\frac{d}{dt} Q(u) \leq -\Lambda(u) + C_4 \Lambda(u) V(u) \leq -\frac{1}{2} \Lambda(u), \tag{3.36}$$

provided that the total variation remains suitably small. Observe that V, Q both remain constant at times where two small h-shocks join together. The previous inequalities together imply

$$\frac{d}{dt} \Big[V(u) + C_1 \cdot Q(u) \Big] \leq 0 \tag{3.37}$$

for some constant C_1, as in (1.9). In particular, if $|u^+ - u^-|$ and η are sufficiently small, then the domain $\mathcal{D}$ in (2.8) is positively invariant.

We now consider the linearized system of equations for a generalized tangent vector $(v, \xi) \in \mathbf{L}^1(\mathbb{R}; \mathbb{R}^n) \times \mathbb{R}^N$, with $N = \#(\mathcal{S} \cup \mathcal{S}')$. Call $v_i = \langle l_i, v \rangle$ the i-th component of v. Outside the jumps, one has

$$
\begin{aligned}
(v_i)_t + \big(\lambda_i^{(h)}(u)v_i\big)_x &= \sum_{j\neq k}\Big\{(\nabla\lambda_i^{(h)}\cdot r_k)(u_x^k v_i - u_x^i v_k) \\
&\qquad + (\lambda_j^{(h)} - \lambda_i^{(h)})\langle l_i,\ [r_j, r_k]\rangle u_x^j v_k\Big\} \\
&\doteq \sum_{j\neq k} H_{ijk}(u)u_x^j v_k.
\end{aligned}
\tag{3.38}
$$

At a point of shock y_α, $\alpha\in\mathcal{S}\cup\mathcal{S}'$, the $(n-1)$ equations (3.12) are satisfied with

$$
w_i^- = v_i^- + \xi_\alpha u_x^{i-}, \qquad w_i^+ = v_i^+ + \xi_\alpha u_x^{i+}.
\tag{3.39}
$$

At the points $y_\alpha^*, y_\alpha^{**}$ where only the h-characteristic speed is discontinuous, we have

$$
v_h^- + \xi_\alpha u_x^{h-} = v_h^+ + \xi_\alpha u_x^{h+}, \qquad v_i^- = v_i^+ \quad (i\neq h).
\tag{3.40}
$$

Recalling (3.28), from (3.40) we obtain a relation between v_h^- and v_h^+. Finally,

$$
\dot\xi_\alpha = D\lambda_{k_\alpha}^{(h)}\big(u(y_\alpha-), u(y_\alpha+)\big)\cdot\Big(\sum_i (v_i^- + \xi_\alpha u_x^{i-})r_i^-,\ \sum_i (v_i^+ + \xi_\alpha u_x^{i+})r_i^+\Big).
\tag{3.41}
$$

Here $\lambda_{k_\alpha}^{(h)}(u^-, u^+)$ is the speed of an ε-approximate shock, as defined at (2.4), while the differential $D\lambda_{k_\alpha}^{(h)}$ is defined as in (1.30). We recall that if $\alpha\in\mathcal{S}$, then the shock speed $\lambda_{k_\alpha}^{(h)}$ is determined by (2.4). In the case $\alpha\in\mathcal{S}'$, the speed $\lambda_{k_\alpha}^{(h)}$ is given by the right hand side of (3.9).

We can now introduce the weighted norm

$$
\big\|(v,\xi)\big\|_u^* \doteq \sum_{i=1}^n \int_{-\infty}^{\infty} W_i^u(x)\big|v_i(x)\big|\,dx + \sum_{\alpha\in\mathcal{S}\cup\mathcal{S}'} W_{k_\alpha}^u(y_\alpha)|\sigma_\alpha|\cdot|\xi_\alpha|,
\tag{3.42}
$$

where the weight functions W_i^u are defined as

$$
W_i^u(x) \doteq 1 + \kappa_1 R_i^u(x) + \kappa_1\kappa_2 Q(u)
\tag{3.43}
$$

for some constants κ_1, κ_2 whose precise value will be determined later. Here $Q(u)$ is the interaction potential (3.32), while $R_i^u(x)$ measures the total amount of waves which approach an i-wave located at x. More precisely, for a point x not coinciding with a large shock, we set

(3.44)

$$
\begin{aligned}
R_i^u(x) \doteq{}& \left[\sum_{j<i}\int_x^{\infty} + \sum_{j>i}\int_{-\infty}^{x}\right]\big|u_x^j(y)\big|\,dy + \left[\sum_{\substack{\alpha\in\mathcal{S}\cup\mathcal{S}'\\ k_\alpha<i,\, y_\alpha>x}} + \sum_{\substack{\alpha\in\mathcal{S}\cup\mathcal{S}'\\ k_\alpha>i,\, y_\alpha<x}}\right]|\sigma_\alpha| \\
&+ \sum_{k_\alpha=i\in\mathcal{S}}|\sigma_\alpha| + \left[\sum_{\substack{\alpha\in\mathcal{S}\\ k_\alpha\leq i,\, y_\alpha^*>x}} + \sum_{\substack{\alpha\in\mathcal{S}\\ k_\alpha>i,\, y_\alpha^*<x}} + \sum_{\substack{\alpha\in\mathcal{S}\\ k_\alpha<i,\, y_\alpha^{**}>x}} + \sum_{\substack{\alpha\in\mathcal{S}\\ k_\alpha\geq i,\, y_\alpha^{**}<x}}\right]\varepsilon.
\end{aligned}
$$

Of course, the third summation in (3.44) contains at most one term: the strength of the i-th big shock, if $i\in\mathcal{S}$. The last summations take care of the fictitious wave-fronts at $y_\alpha^*, y_\alpha^{**}$, regarded as waves of the α-th family, of strength ε. The presence of these terms takes into account the fact that some wave speeds may experience a small change across these lines. In the case $x = y_\alpha$ for some $\alpha \in \mathcal{S}$ we set

$$
\begin{aligned}
R^u_{k_\alpha}(y_\alpha) \doteq & \left[\sum_{j\le k_\alpha}\int_{y_\alpha}^{\infty} + \sum_{j\ge k_\alpha}\int_{-\infty}^{y_\alpha}\right] |u_x^j(x)|\, dx \\
& + \left[\sum_{\substack{\beta\in\mathcal{S}\cup\mathcal{S}'\\ k_\beta\le k_\alpha,\, y_\beta>y_\alpha}} + \sum_{\substack{\beta\in\mathcal{S}\cup\mathcal{S}'\\ k_\beta\ge k_\alpha,\, y_\beta<y_\alpha}}\right] |\sigma_\beta| \\
& + \varepsilon\cdot\left\{\int_{y_\alpha^*}^{y_\alpha}\Big(\sum_{j<k_\alpha}|u_x^j(x)| - \sum_{j\ge k_\alpha}|u_x^j(x)|\Big)dx\right. \\
& + \int_{y_\alpha}^{y_\alpha^{**}}\Big(\sum_{j>k_\alpha}|u_x^j(x)| - \sum_{j\le k_\alpha}|u_x^j(x)|\Big)dx \\
& \left. + \sum_{\substack{\beta\in\mathcal{S}'\\ y_\beta\in[y_\alpha^*,y_\alpha^{**}]}} |\sigma_\beta|\text{sign}\,\big[(\dot y_\beta-\dot y_\alpha)(y_\beta-y_\alpha)\big]\right\}.
\end{aligned}
\tag{3.45}
$$

Following [**B4**], we now observe that the derivative of R_i along an i-characteristic satisfies the estimate

$$
(R_i^u)_t + \lambda_i^{(h)}(u)(R_i^u)_x \le -\sum_{j\ne i}|\lambda_j^{(h)}(u) - \lambda_i^{(h)}(u)||u_x^j| + C\Lambda(u).
\tag{3.46}
$$

Moreover, recalling (3.35), along a shock we have

$$
\begin{aligned}
\frac{d}{dt}R^u_{k_\alpha}(y_\alpha(t)) \le & -\sum_{j\le k_\alpha}(\dot y_\alpha - \lambda_j^{(h)}(y_\alpha+))|u_x^{j+}| \\
& -\sum_{j\ge k_\alpha}(\lambda_j^{(h)}(y_\alpha-) - \dot y_\alpha)|u_x^{j-}| \\
& + C\Lambda(u) - \varepsilon\Big(\sum_j|\lambda_j^{(h)}(y_\alpha^*+) - \dot y_\alpha||u_x^j(y_\alpha^*+)| \\
& + \sum_j|\lambda_j^{(h)}(y_\alpha^{**}-) - \dot y_\alpha||u_x^j(y_\alpha^{**}-)|\Big) + \frac{\varepsilon}{|\sigma_\alpha|}C\Lambda_\alpha(u) \\
\le & -\frac{\Lambda_\alpha(u)}{2|\sigma_\alpha|} + C\Lambda(u) - \varepsilon\Lambda_\alpha^*(u)
\end{aligned}
\tag{3.47}
$$

in case $\alpha \in \mathcal{S}$. We use here the notation

$$
\begin{aligned}
\Lambda_\alpha^*(u) \doteq & \sum_j|\lambda_j^{(h)}(y_\alpha^*+) - \dot y_\alpha||u_x^j(y_\alpha^*+)| \\
& + \sum_j|\lambda_j^{(h)}(y_\alpha^{**}-) - \dot y_\alpha||u_x^j(y_\alpha^{**}-)|.
\end{aligned}
\tag{3.48}
$$

Since the constant C is independent of ε and we are eventually interested in the limit $\varepsilon \to 0$, in (3.47) it was not restrictive to assume $\varepsilon C < 1/2$. When $\alpha \in \mathcal{S}'$ we have the simpler estimate

$$
\begin{aligned}
\frac{d}{dt} R^u_{k_\alpha}(y_\alpha(t)) \leq & - \sum_{j<k_\alpha} \left(\dot{y}_\alpha - \lambda_j^{(h)}(y_\alpha+)\right)|u_x^{j+}| \\
& - \sum_{j>k_\alpha} \left(\lambda_j^{(h)}(y_\alpha-) - \dot{y}_\alpha\right)|u_x^{j-}| + C\Lambda(u) \\
\leq & - \frac{\Lambda_\alpha(u)}{|\sigma_\alpha|} + C\Lambda(u).
\end{aligned} \tag{3.49}
$$

The time derivative of the weighted norm of a tangent vector can now be computed as

$$
\begin{aligned}
\frac{d}{dt}\left\|(v(t),\xi(t))\right\|^*_{u(t)} \leq & \sum_{\alpha\in\mathcal{S}\cup\mathcal{S}'} \Big[(-\dot{\sigma}_\alpha)W^u_{k_\alpha}(y_\alpha)|\xi_\alpha| + |\sigma_\alpha|\dot{W}^u_{k_\alpha}(y_\alpha)|\xi_\alpha| \\
& \qquad + |\sigma_\alpha|W^u_{k_\alpha}(y_\alpha)(\text{sign }\xi_\alpha)\dot{\xi}_\alpha\Big] \\
& + \sum_{i=1}^n \int_{-\infty}^{\infty} \left[(W_i^u)_t + \lambda_i^{(h)}(u)(W_i^u)_x\right]|v_i|\,dx \\
& + \sum_{i=1}^n \int_{-\infty}^{\infty} W_i^u \cdot (\text{sign }v_i)\Big((v_i)_t + \left(\lambda_i^{(h)}(u)v_i\right)_x\Big)\,dx \\
& + \sum_{\alpha\in\mathcal{S}\cup\mathcal{S}'}\sum_{i=1}^n \Big[W_i^u(y_\alpha+)\left(\lambda_i^{(h)}(y_\alpha+) - \dot{y}_\alpha\right)\left|v_i(y_\alpha+)\right| \\
& \qquad - W_i^u(y_\alpha-)\left(\lambda_i^{(h)}(y_\alpha-) - \dot{y}_\alpha\right)\left|v_i(y_\alpha-)\right|\Big] \\
& + \sum_{\alpha\in\mathcal{S}}\sum_{i=1}^n \Big[W_i^u(y^*_\alpha+)\left(\lambda_i^{(h)}(y^*_\alpha+) - \dot{y}_\alpha\right)\left|v_i(y^*_\alpha+)\right| \\
& \qquad - W_i^u(y^*_\alpha-)\left(\lambda_i^{(h)}(y^*_\alpha-) - \dot{y}_\alpha\right)\left|v_i(y^*_\alpha-)\right|\Big] \\
& + \sum_{\alpha\in\mathcal{S}}\sum_{i=1}^n \Big[W_i^u(y^{**}_\alpha+)\left(\lambda_i^{(h)}(y^{**}_\alpha+) - \dot{y}_\alpha\right)\left|v_i(y^{**}_\alpha+)\right| \\
& \qquad - W_i^u(y^{**}_\alpha-)\left(\lambda_i^{(h)}(y^{**}_\alpha-) - \dot{y}_\alpha\right)\left|v_i(y^{**}_\alpha-)\right|\Big] \\
\doteq & E_1 + E_2 + E_3 + E_4 + E_5 + E_6.
\end{aligned} \tag{3.50}
$$

We claim that the right hand side of (3.50) is non-positive, as long as u remains in the domain $\mathcal{D}$ in (2.8), for η suitably small. Before embarking in the lengthy computations that follow, the reader is advised to review the Example 1.6 given in the Introduction. In the scalar case, thanks to the identities (1.41)–(1.44), one can choose the weight function $W^u \equiv 1$ and obtain (1.45) by a straightforward computation. To handle the vector valued case, we try to use a similar argument for each component of the tangent vector. More precisely, we replace (1.41) by (3.38), (1.42) by (3.41), (1.43) by (3.29), and (1.44) by (3.59). In contrast with the scalar case, the relations (3.38), (3.29) and (3.59) now contain a non-zero right hand side. The key point is that all these extra terms are due to some kind of interaction. Therefore, their contribution to the growth of the norm $\left\|(v,\xi)\right\|^*_u$ can be more than compensated by the decrease of suitable weight functions.

The estimation of (3.50) follows [**B4**], with suitable modifications. By choosing η sufficiently small, we can assume that all the quantities

$$\begin{gathered}\big|u(y^*_\alpha)-\omega_{\alpha-1}\big|,\ \big|u(y_\alpha-)-\omega_{\alpha-1}\big|,\ \big|u(y_\alpha+)-\omega_\alpha\big|,\ \big|u(y^{**}_\alpha)-\omega_\alpha\big|,\\ \big|\lambda^{(h)}_h(y^*_\alpha+)-\lambda^{(h)}_h(y^*_\alpha-)\big|,\ \big|\lambda^{(h)}_h(y^{**}_\alpha+)-\lambda^{(h)}_h(y^{**}_\alpha-)\big|,\end{gathered}\tag{3.51}$$

are as small as we like. Define the *instantaneous amount of interaction* between u and v by setting

$$\Psi(u,v)\doteq\tilde\Psi(u,v)+\sum_{\alpha\in\mathcal{S}\cup\mathcal{S}'}\Psi_\alpha(u,v),\tag{3.52}$$

$$\tilde\Psi(u,v)\doteq\sum_{i\neq j}\int_{-\infty}^{\infty}\big|\lambda^{(h)}_i(u)-\lambda^{(h)}_j(u)\big|\big|u^i_x(x)\big|\big|v_j(x)\big|\,dx,\tag{3.53}$$

$$\Psi_\alpha(u,v)\doteq\begin{cases}|\sigma_\alpha|\Big[\sum_{i\geq k_\alpha}\big(\lambda^{(h)}_i(y_\alpha-)-\dot y_\alpha\big)\big|v_i(y_\alpha-)\big| \\ \qquad+\sum_{i\leq k_\alpha}\big(\dot y_\alpha-\lambda^{(h)}_i(y_\alpha+)\big)\big|v_i(y_\alpha+)\big|\Big] & \text{if } \alpha\in\mathcal{S},\\ |\sigma_\alpha|\Big[\sum_{i> k_\alpha}\big(\lambda^{(h)}_i(y_\alpha-)-\dot y_\alpha\big)\big|v_i(y_\alpha-)\big| \\ \qquad+\sum_{i< k_\alpha}\big(\dot y_\alpha-\lambda^{(h)}_i(y_\alpha+)\big)\big|v_i(y_\alpha+)\big|\Big] & \text{if } \alpha\in\mathcal{S}'.\end{cases}\tag{3.54}$$

Define also the upper bound for all weights

$$M^u_W\doteq\sup\big\{W^u_i(x);\,x\in\mathbb{R},\ i=1,\dots,n\big\}.$$

By (3.46), (3.36), the time derivative of W^u_i along an i-characteristic satisfies

$$\begin{aligned}\left(\frac{\partial}{\partial t}+\lambda^{(h)}_i(u)\frac{\partial}{\partial x}\right)W^u_i\leq&-\kappa_1\sum_{j\neq i}\big|\lambda^{(h)}_j(u)-\lambda^{(h)}_i(u)\big|\big|u^j_x\big|\\ &+\kappa_1 C\Lambda(u)+\kappa_1\kappa_2\dot Q(u)\\ \leq&-\kappa_1\sum_{j\neq i}\big|\lambda^{(h)}_j(u)-\lambda^{(h)}_i(u)\big|\big|u^j_x\big|,\end{aligned}\tag{3.55}$$

provided that κ_2 is chosen sufficiently large. Along a shock, by (3.47) and (3.36) one has

$$\begin{aligned}\frac{d}{dt}W^u_{k_\alpha}\big(y_\alpha(t)\big)&\leq-\kappa_1\frac{\Lambda_\alpha(u)}{2|\sigma_\alpha|}-\kappa_1\varepsilon\Lambda^*_\alpha(u)+\kappa_1\big(C\Lambda(u)+\kappa_2\dot Q(u)\big)\\ &\leq-\kappa_1\frac{\Lambda_\alpha(u)}{2|\sigma_\alpha|}-\kappa_1\varepsilon\Lambda^*_\alpha(u)\end{aligned}\tag{3.56}$$

if $\alpha\in\mathcal{S}$. In the case $\alpha\in\mathcal{S}'$, we have the simpler estimate

$$\frac{d}{dt}W^u_{k_\alpha}\big(y_\alpha(t)\big)\leq-\kappa_1\frac{\Lambda_\alpha(u)}{|\sigma_\alpha|}+\kappa_1\big(C\Lambda(u)+\kappa_2\dot Q(u)\big)\leq-\kappa_1\frac{\Lambda_\alpha(u)}{|\sigma_\alpha|}.\tag{3.57}$$

We now observe that for every $\alpha\in\mathcal{S}\cup\mathcal{S}'$ there holds

$$\sum_{i\neq k_\alpha}\big|v_i(y_\alpha+)-v_i(y_\alpha-)\big|\leq C\cdot\big(|\xi_\alpha|\Lambda_\alpha(u)+\Psi_\alpha(u,v)\big).\tag{3.58}$$

Moreover, for $\alpha \in \mathcal{S}$ and any $\eta^-, \eta^+ \in \mathbb{R}$ we have

$$\begin{aligned}\Big|\big(\lambda_{k_\alpha}^{(h)}(y_\alpha-) - \dot y_\alpha\big)\eta^- + \big(\dot y_\alpha - \lambda_{k_\alpha}^{(h)}(y_\alpha+)\big)\eta^+ \\ - |\sigma_\alpha| \cdot D\lambda_{k_\alpha}^{(h)}\big(u(y_\alpha-), u(y_\alpha+)\big) \cdot \big(\eta^- r_{k_\alpha}^-,\ \eta^+ r_{k_\alpha}^+\big)\Big| \\ \leq C \cdot |\sigma_\alpha|^2\big(|\eta^-| + |\eta^+|\big)\end{aligned} \tag{3.59}$$

for some constant C. In the case $\alpha \in \mathcal{S}'$ the same result holds with $C = 0$. These estimates should be compared with the identity (1.44), valid in the scalar case.

We are now ready to estimate each term on the right hand side of (3.50). Using (3.29), (3.56), (3.57), (3.41), then (3.59) with $\eta^\pm = u_x^{k_\alpha \pm}, v_{k_\alpha}^\pm$, we obtain

$$\begin{aligned}E_1 \leq & - \sum_{\alpha \in \mathcal{S}\cup\mathcal{S}'} \Big[\big(\lambda_{k_\alpha}^{(h)}(y_\alpha-) - \dot y_\alpha\big)u_x^{k_\alpha-} + \big(\dot y_\alpha - \lambda_{k_\alpha}^{(h)}(y_\alpha+)\big)u_x^{k_\alpha+}\Big] W_{k_\alpha}^u(y_\alpha)|\xi_\alpha| \\ & + C\Big(\sum_{\alpha\in\mathcal{S}} |\sigma_\alpha|^3\big(|u_x^{k_\alpha+}| + |u_x^{k_\alpha-}|\big) \\ & \qquad + \sum_{\alpha \in \mathcal{S}\cup\mathcal{S}'} |\sigma_\alpha| \sum_{i\neq k_\alpha} \big(|u_x^{i+}| + |u_x^{i-}|\big)\Big) W_{k_\alpha}^u(y_\alpha)|\xi_\alpha| \\ & - \frac{\kappa_1}{2} \sum_{\alpha \in \mathcal{S}\cup\mathcal{S}'} |\xi_\alpha|\Lambda_\alpha(u) - \kappa_1\varepsilon \sum_{\alpha\in\mathcal{S}} |\xi_\alpha|\Lambda_\alpha^*(u) \\ & + \sum_{\alpha \in \mathcal{S}\cup\mathcal{S}'} |\sigma_\alpha| W_{k_\alpha}^u(y_\alpha)(\text{sign }\xi_\alpha) \cdot D\lambda_{k_\alpha}^{(h)}\big(u(y_\alpha-), u(y_\alpha+)\big) \cdot \\ & \qquad \cdot \Big(\sum_i (v_i^- + \xi_\alpha u_x^{i-})r_i^-,\ \sum_i (v_i^+ + \xi_\alpha u_x^{i+})r_i^+\Big) \\ \leq & - \frac{\kappa_1}{2} \sum_{\alpha \in \mathcal{S}\cup\mathcal{S}'} |\xi_\alpha|\Lambda_\alpha(u) - \kappa_1\varepsilon \sum_{\alpha\in\mathcal{S}} |\xi_\alpha|\Lambda_\alpha^*(u) \\ & + \sum_{\alpha \in \mathcal{S}\cup\mathcal{S}'} \Big[\big(\lambda_{k_\alpha}^{(h)}(y_\alpha-) - \dot y_\alpha\big)v_{k_\alpha}^- + \big(\dot y_\alpha - \lambda_{k_\alpha}^{(h)}(y_\alpha+)\big)v_{k_\alpha}^+\Big](\text{sign }\xi_\alpha)W_{k_\alpha}^u(y_\alpha)|\xi_\alpha| \\ & + \sum_{\alpha \in \mathcal{S}\cup\mathcal{S}'} CM_W^u\Big(\Psi_\alpha(u,v) + |\xi_\alpha|\Lambda_\alpha(u)\Big).\end{aligned} \tag{3.60}$$

By (3.55), (3.38) and strict hyperbolicity, the second and third term in (3.50) satisfy

$$E_2 \leq -\kappa_1 \tilde\Psi(u,v) \tag{3.61}$$

$$E_3 \leq CM_W^u \tilde\Psi(u,v). \tag{3.62}$$

By (3.54) and (3.58), the fourth term is estimated by

$$\begin{aligned}E_4 \leq & - \kappa_1 \sum_{\alpha \in \mathcal{S}\cup\mathcal{S}'} \Psi_\alpha(u,v) + CM_W^u \sum_{\alpha \in \mathcal{S}\cup\mathcal{S}'} \Big(\Psi_\alpha(u,v) + |\xi_\alpha|\Lambda_\alpha(u)\Big) \\ & + \sum_{\alpha \in \mathcal{S}\cup\mathcal{S}'} \Big[\big(\lambda_{k_\alpha}^{(h)}(y_\alpha+) - \dot y_\alpha\big)\big|v_{k_\alpha}(y_\alpha+)\big| \\ & \qquad - \big(\lambda_{k_\alpha}^{(h)}(y_\alpha-) - \dot y_\alpha\big)\big|v_{k_\alpha}(y_\alpha-)\big|\Big] W_{k_\alpha}^u(y_\alpha).\end{aligned} \tag{3.63}$$

Next, consider any term in the summation E_5. From the definition (3.45) it follows

$$W_i^u(y_\alpha^*+) = W_i^u(y_\alpha^*-) - \varepsilon \cdot \text{sign}\,\big(\lambda_i^{(h)}(y_\alpha^*-) - \dot y_\alpha\big). \tag{3.64}$$

If $i \neq h$, we simply have

$$v_i(y_\alpha^*+) = v_i(y_\alpha^*-), \qquad \lambda_i^{(h)}(y_\alpha^*+) = \lambda_i^{(h)}(y_\alpha^*-),$$

hence

$$\begin{aligned}\big(\lambda_i^{(h)}(y_\alpha^*+)-\dot y_\alpha\big)&W_i^u(y_\alpha^*+)\big|v_i(y_\alpha^*+)\big| \\ &- \big(\lambda_i^{(h)}(y_\alpha^*-) - \dot y_\alpha\big)W_i^u(y_\alpha^*-)\big|v_i(y_\alpha^*-)\big| \\ &= -\varepsilon\Big|\lambda_i^{(h)}(y_\alpha^*) - \dot y_\alpha\Big|\big|v_i(y_\alpha^*)\big|.\end{aligned} \tag{3.65}$$

In the case $i = h$, recalling (3.28) and (3.40), for any $\varepsilon' > 0$, by choosing η small in (2.8) we can achieve the estimates

$$\begin{aligned}\big|\lambda_h^{(h)}(y_\alpha^*+) - \lambda_h^{(h)}(y_\alpha^*-)\big| &\le \varepsilon' < \varepsilon^3, \\ \big|v_h(y_\alpha^*+) - v_h(y_\alpha^*-)\big| = |\xi_\alpha|\big|u_x^{h+} - u_x^{h-}\big| &\le \varepsilon'|\xi_\alpha| \cdot \min\big\{|u_x^{h+}|,\ |u_x^{h-}|\big\},\end{aligned} \tag{3.66}$$

$$\begin{aligned}\big|\lambda_h^{(h)}(y_\alpha^*\pm) - \dot y_\alpha\big| &> \frac{1}{2}\big|\lambda_h(\omega_{\alpha-1}) - \dot y_\alpha\big| > c\varepsilon > 0, \\ \big|\lambda_h^{(h)}(y_\alpha^{**}\pm) - \dot y_\alpha\big| &> \frac{1}{2}\big|\lambda_h(\omega_\alpha) - \dot y_\alpha\big| > c\varepsilon > 0,\end{aligned} \tag{3.67}$$

for some constant $c > 0$. To fix the ideas, assume $h \ge k_\alpha$. The bounds (3.66), (3.67) with ε' sufficiently small imply

$$\begin{aligned}\big(\lambda_h^{(h)}&(y_\alpha^*+) - \dot y_\alpha\big)W_h^u(y_\alpha^*+)\big|v_h(y_\alpha^*+)\big| \\ &- \big(\lambda_h^{(h)}(y_\alpha^*-) - \dot y_\alpha\big)W_h^u(y_\alpha^*-)\big|v_h(y_\alpha^*-)\big| \\ &= \big(\lambda_h^{(h)}(y_\alpha^*+) - \dot y_\alpha\big)W_h^u(y_\alpha^*+)\Big(\big|v_h(y_\alpha^*+)\big| - \big|v_h(y_\alpha^*-)\big|\Big) \\ &+ \big(\lambda_h^{(h)}(y_\alpha^*+) - \lambda_h^{(h)}(y_\alpha^*-)\big)W_h^u(y_\alpha^*+)\big|v_h(y_\alpha^*-)\big| \\ &+ \big(\lambda_h^{(h)}(y_\alpha^*-) - \dot y_\alpha\big)\big(W_h^u(y_\alpha^*+) - W_h^u(y_\alpha^*-)\big)\big|v_h(y_\alpha^*-)\big| \\ &\le \big|\lambda_h^{(h)}(y_\alpha^*+) - \dot y_\alpha\big|M_W^u\varepsilon'|\xi_\alpha|\big|u_x^h(y_\alpha^*+)\big| \\ &\quad + \varepsilon' M_W^u\big|v_h(y_\alpha^*-)\big| - \varepsilon\big|\lambda_h^{(h)}(y_\alpha^*-) - \dot y_\alpha\big|\big|v_h(y_\alpha^*-)\big| \\ &\le \big|\lambda_h^{(h)}(y_\alpha^*+) - \dot y_\alpha\big|M_W^u\varepsilon'|\xi_\alpha|\big|u_x^h(y_\alpha^*+)\big|.\end{aligned} \tag{3.68}$$

Each term in the summations for E_5 and E_6 can be estimated in the same way. Recalling the definition (3.48), we thus have

$$E_5 + E_6 \le \varepsilon' M_W^u \sum_{\alpha\in S} |\xi_\alpha|\Lambda_\alpha^*(u). \tag{3.69}$$

From (3.50), combining the estimates (3.60)–(3.63) and (3.69) we finally obtain

$$
\begin{aligned}
E_1 + \cdots + E_6 \leq & -\frac{\kappa_1}{2} \sum_{\alpha \in \mathcal{S} \cup \mathcal{S}'} |\xi_\alpha| \Lambda_\alpha(u) - \kappa_1 \varepsilon \sum_{\alpha \in \mathcal{S}} |\xi_\alpha| \Lambda_\alpha^*(u) \\
& + C M_W^u \sum_{\alpha \in \mathcal{S} \cup \mathcal{S}'} \Big(\Psi_\alpha(u,v) + |\xi_\alpha| \Lambda_\alpha(u) \Big) \\
& - \kappa_1 \tilde{\Psi}(u,v) + C M_W^u \tilde{\Psi}(u,v) \\
& - \kappa_1 \sum_{\alpha \in \mathcal{S} \cup \mathcal{S}'} \Psi_\alpha(u,v) + \varepsilon' M_W^u \sum_{\alpha \in \mathcal{S}} |\xi_\alpha| \Lambda_\alpha^*(u).
\end{aligned}
\tag{3.70}
$$

We now choose $\kappa_1 = 4C$, then let the total variation be small enough so that $M_W^u \leq 2$, and finally choose $\varepsilon', \eta > 0$ so small that $\varepsilon' M_W^u \leq \kappa_1 \varepsilon$. These choices imply

$$
\frac{d}{dt} \Big\| \big(v(t), \xi(t)\big) \Big\|_{u(t)}^* \leq 0,
\tag{3.71}
$$

at every time t where no shock interaction occurs and no small h-shock crosses one of the lines $y_\alpha^*, y_\alpha^{**}$, $\alpha \in \mathcal{S}$.

Next, we show that the weighted norm (3.42) decreases at every time τ of interaction. Let y', y'' be the locations of two h-shocks, let $\sigma', \sigma'' < 0$ be their strengths and let ξ', ξ'' be their shift rates, before the interaction time τ. Call y, σ, ξ the corresponding quantities after the interaction. Since both shocks are small, by the coincidence of shock and rarefaction curves it follows

$$
\sigma(\tau+) = \sigma'(\tau-) + \sigma''(\tau-), \qquad \xi(\tau+) = \frac{\sigma'(\tau-)\xi'(\tau-) + \sigma''(\tau-)\xi''(\tau-)}{\sigma'(\tau-) + \sigma''(\tau-)}.
\tag{3.72}
$$

Moreover, since all weight functions decrease, we have

$$
W_h^u\big(y(\tau+)\big) < \min\Big\{ W_h^u\big(y'(\tau-)\big),\ W_h^u\big(y''(\tau-)\big) \Big\},
\tag{3.73}
$$

$$
\Big\| (v,\xi)(\tau+) \Big\|_{u(\tau+)}^* < \Big\| (v,\xi)(\tau-) \Big\|_{u(\tau-)}^*.
\tag{3.74}
$$

We do not need to consider the case where three or more small shocks interact at the same point, because this situation is non-generic, and can be avoided by an arbitrarily small perturbation of the evolution equations. More precisely, let $\varepsilon^* > 0$ be given, together with a regular path $\theta \mapsto u^\theta(t_0)$, at some time t_0. Then we can find a smooth function φ, with $\|\varphi\|_{C^3} < \varepsilon^*$, such that, if we replace the h-wave speed $\lambda_h^{(h)}$ in (3.6) with the slightly perturbed value

$$
\lambda_h^\dagger \doteq \lambda_h^{(h)} + \varphi(t, x, \theta),
$$

then, for all but finitely many θ, the corresponding approximate solution u^θ, for $t \geq t_0$, has shocks interacting only two at a time.

Finally, consider the case of a small h-shock which enters (or exits from) a shock layer. To fix the ideas, assume that the small h-shock crosses the line y_α^* from left to right, thus entering the shock layer around a large shock at y_α. Observe that the sizes of the two shocks satisfy

$$
\sigma_\alpha < -\varepsilon, \qquad -3\varepsilon < \sigma_\beta < 0
\tag{3.75}
$$

and do not change at the time τ when the crossing occurs. Call ξ_α the shift rate of the large shock at y_α (clearly, y_α^*, y_α^{**} shift at exactly the same rate), and let ξ_β^-, ξ_β^+ be the shift rates of the small h-shock before and after the crossing. An elementary computation yields

$$\xi_\beta^+ = \frac{\xi_\alpha(\dot y_\beta^- - \dot y_\beta^+) - \xi_\beta^-(\dot y_\alpha - \dot y_\beta^+)}{\dot y_\beta^- - \dot y_\alpha} = \xi_\beta^- + \frac{(\xi_\beta^- - \xi_\alpha)(\dot y_\beta^+ - \dot y_\beta^-)}{\dot y_\beta^- - \dot y_\alpha}, \tag{3.76}$$

where $\dot y_\beta^-, \dot y_\beta^+$ are the speeds of the small shock before and after the crossing, respectively. At the time of crossing, the change in the weighted norm of the generalized tangent vector is now computed by

$$\begin{aligned} \Big\|(v,\xi)(\tau+)\Big\|^*_{u(\tau+)} - \Big\|(v,\xi)(\tau-)\Big\|^*_{u(\tau-)} &= \\ =&|\sigma_\beta||\xi_\beta^+|W_h^+(y_\beta) + |\sigma_\alpha||\xi_\alpha|W_{k_\alpha}^+(y_\alpha) \\ &- |\sigma_\beta||\xi_\beta^-|W_h^-(y_\beta) - |\sigma_\alpha||\xi_\alpha|W_{k_\alpha}^-(y_\alpha). \end{aligned} \tag{3.77}$$

From the definitions (3.43)–(3.45) it follows

$$W_h^+(y_\beta) - W_h^-(y_\beta) = -\kappa_1\varepsilon, \qquad W_{k_\alpha}^+(y_\alpha) - W_{k_\alpha}^-(y_\alpha) = -\kappa_1\varepsilon|\sigma_\beta|. \tag{3.78}$$

Moreover, by (3.66)-(3.67) we have

$$|\dot y_\beta^+ - \dot y_\beta^-| \le \varepsilon', \qquad |\dot y_\beta^- - \dot y_\alpha| > c\varepsilon. \tag{3.79}$$

Therefore, if ε' is sufficiently small (which can be accomplished by choosing η small in (2.8)), by (3.78), (3.79) and (3.76) we again obtain (3.74).

By (3.71) and (3.74), the weighted length $\|\gamma\|_*$ of a path of approximate ε-solutions $\gamma_t\colon \theta \mapsto u^\theta(t,\cdot)$ does not increase in time, as long as all functions u^θ retain their regularity, remaining piecewise Lipschitz continuous.

REMARK 3.1. In order to obtain the previous estimates, we used the fact that the amount of waves inside a shock layer $[y_\alpha^*, y_\alpha^{**}] \setminus \{y_\alpha\}$ is small compared with the strength of the shock at y_α. More precisely, both quantities

[strength of waves inside the shock layer],

[change in the characteristic speed $\lambda_h^{(h)}$ across the boundaries $y_\alpha^*, y_\alpha^{**}$]

should be $<<$ [strength of the shock at y_α]^2. This motivates the choices of the exponents in Cases 2-3 of Definition 2.6, in Section 2.

CHAPTER 4

Restarting Procedures

In this section we complete the construction of the local semigroups, describing the restarting procedures and carefully estimating how the various weighted norms are changed at restarting times. For clarity of exposition, we shall write the statements of the various lemmas one after the other, collecting all the proofs at the end of the section.

Let $\delta_1, \delta_2 > 0$ be given, together with a Riemann data (u^-, u^+). As in the previous section, call $\mathcal{S} = \{j_1, \ldots, j_\nu\} \subseteq \{1, \ldots, n\}$ the set of indices i such that the solution of the Riemann problem (2.6)-(1.1) contains an i-shock of strength $|\sigma_i| \geq 2\varepsilon$. Let $\eta > 0$ be chosen small enough, so that every function u in the domain $\mathcal{D}$ at (2.8) contains exactly one i-shock of strength $> \varepsilon$ for each $i \in \mathcal{S}$, while all other jumps in u have strength $< 3\varepsilon$.

We now introduce the domain $\mathcal{D}^{\delta_2} \subset \mathcal{D}$ consisting of all piecewise Lipschitz functions $u \in \mathcal{D}$ such that

- The large shocks of u are located at points y_i, $i \in \mathcal{S}$, with $y_j - y_i > 2\delta_2$ whenever $i < j$.
- u contains no jump inside the set

$$\bigcup_{i \in \mathcal{S}} \Big([y_i - \delta_2/3,\ y_i[\cup]y_i,\ y_i + \delta_2/3] \Big), \tag{4.1}$$

- Outside the set

$$\bigcup_{i \in \mathcal{S}} [y_i - \delta_2/3,\ y_i + \delta_2/3], \tag{4.2}$$

all jumps in u belong to a single characteristic family.

DEFINITION 4.1. For a given $\delta > 0$, we say that a function $u\colon [0, +\infty) \to \mathcal{D}^{\delta_2}$ is a δ-accurate approximate ε-solution of (1.1) if the following holds.

(1) u is continuous with values in $\mathbf{L}^1$, except for a countable set of times $\tau_1 < \tau_2 < \ldots$, with $\tau_i \to +\infty$.

(2) Each interval $J_\ell \doteq [\tau_\ell, \tau_{\ell+1}[$ is contained in some $I_{m,h}$, defined at (2.9). For $t \in J_\ell \subseteq I_{m,h}$, the function u satisfies the quasilinear hyperbolic equation (3.8) outside the shocks, the conditions (2.3)-(2.4) along the large shocks and the conditions (3.3), (3.9) along the small h-shocks.

(3) For every $T > 0$, the restartings performed at the times τ_ℓ satisfy

$$\sum_{\tau_\ell \leq T} \big\| u(\tau_\ell+) - u(\tau_\ell-) \big\|_{\mathbf{L}^1} \leq \delta T. \tag{4.3}$$

The construction of a δ-accurate approximate ε-solution can be achieved following Section 3 and using the three restarting procedures described below. For every function $u \in \mathcal{D}^{\delta_2}$ we denote by Lip(u) its Lipschitz constant, and define the weighted Lipschitz constant

$$\text{w-Lip}(u) \doteq \max_i \sup_x |u_x^i(x)| \exp\left[\beta W_i^u(x)\right], \tag{4.4}$$

where the weights W_i^u are as in (3.43) and β is a suitably large constant, whose precise value will be determined later. For convenience, some additional domains are now defined.

- $\mathcal{D}_{h,L}$ is the set of all functions $u \in \mathcal{D}^{\delta_2}$ whose weighted Lipschitz constant is $\leq L$ and whose shocks, outside the set (4.2), all belong to the the h-th family.
- $\mathcal{D}^*_{h,L}$ is the set of all functions $u \in \mathcal{D}_{h,L}$ which satisfy the additional condition

$$u_x^h(x) \exp\left[\beta W_h^u(x)\right] \geq -1, \tag{4.5}$$

outside the set

$$\bigcup_{i \in \mathcal{S}} \Big([y_i - 2\delta_2/3,\ y_i + 2\delta_2/3] \Big). \tag{4.6}$$

- $\mathcal{D}'_{h,L}$ is the set of all functions $u \in \mathcal{D}_{h,L}$ which are Lipschitz continuous inside the set

$$\bigcup_{i \in \mathcal{S}} \Big([y_i - 2\delta_2/3,\ y_i[\ \cup\ ]y_i,\ y_i + 2\delta_2/3] \Big). \tag{4.7}$$

Given an initial data $\bar{u} \in \mathcal{D}^{\delta_2}$, we now construct a δ-approximate ε-solution u, defined for all $t \in [0, \infty[$, with $u(t) \in \mathcal{D}_{h,L}$ on every $I_{m,h}$ for some weighted Lipschitz constant $L = L(m,h)$.

By induction, assume that an approximate solution has been constructed on the interval $[0,\ \tau_{m,h-1}]$ for some m, h as in (2.9), and assume that $u(\tau_{m,h-1}) \in \mathcal{D}^*_{h,L} \cap \mathcal{D}'_{h,L}$ for some $L > 0$. We shall prolong this solution up to time $\tau_{m,h}$ in such a way that

$$u(t) \in \mathcal{D}_{h,L'} \qquad \forall t \in I_{m,h}, \tag{4.8}$$

$$u(\tau_{m,h}) \in \mathcal{D}^*_{h+1,L''} \cap \mathcal{D}'_{h+1,L''}, \tag{4.9}$$

for some constants L', L'' [if $h = n$, then $u(\tau_{m,h}) \in \mathcal{D}^*_{1,L''} \cap \mathcal{D}'_{1,L''}$]. Iterating this argument, we thus obtain an approximate solution u, defined for all $t \geq 0$.

In the following, we refer to a piecewise Lipschitz solution of the quasilinear system (3.8), satisfying the ε-Rankine-Hugoniot equations (2.3)-(2.4) along the big jumps and (3.3), (3.9) along the small h-shocks. To retain the piecewise Lipschitz regularity, three types of restartings will be needed.

(a) At the terminal time $t = \tau_{m,h}$, we replace a function whose small shocks all belong to the h-family with a new function whose small shocks belong to the $(h+1)$-family [to the 1-family if $h = n$].

(b) When a small h-shock penetrates the set (4.7) around one of the large shocks at y_i, $i \in \mathcal{S}$, the small shock is replaced by a smooth compressive wave. Since all waves travel with speed < 1, to avoid the interaction between the two shocks (more

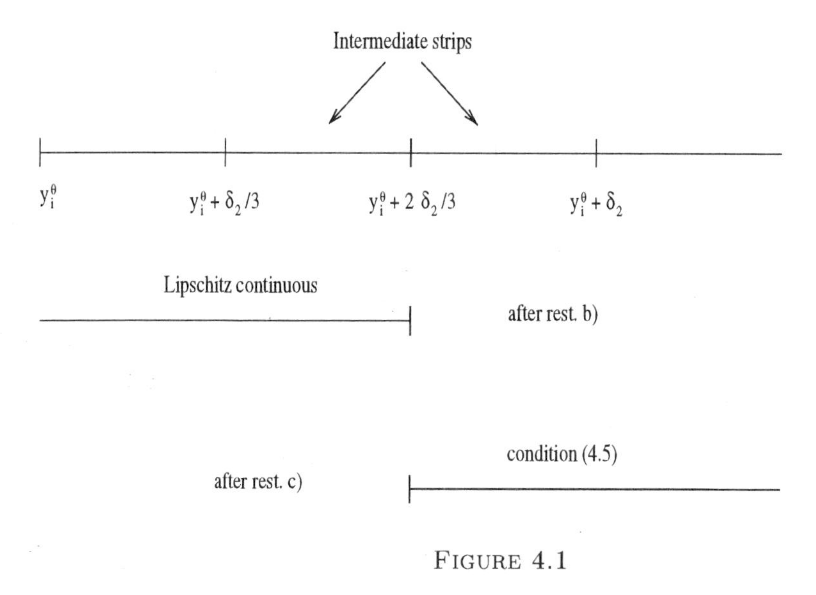

FIGURE 4.1

precisely, to prevent small shocks from entering the set (4.2)), it suffices that we perform these restartings at times

$$\tau_{m,h-1} + \Delta' t, \ \tau_{m,h-1} + 2\Delta' t, \ \ldots \ , \ \tau_{m,h-1} + M\Delta' t = \tau_{m,h}$$

for a suitable time step $\Delta' t < \delta_2/12$. More precisely, calling $[\![s]\!]$ the integer part of s, we choose

$$M \doteq 1 + \left[\!\!\left[\frac{\delta_1/n}{\delta_2/12}\right]\!\!\right], \qquad \Delta' t \doteq \frac{\tau_{m,h} - \tau_{m,h-1}}{M} = \frac{\delta_1}{nM}. \tag{4.10}$$

(c) When the (genuinely nonlinear) gradient component u_x^h becomes too large and negative and a gradient catastrophe is about to occur, we replace steep compressive h-waves with several small h-shocks. In this way, after the restarting, the new function will satisfy $u_x^h(x) \exp\left[\beta W_h^u(x)\right] \geq -1$ outside the set (4.6).

The two properties:

- $u(t)$ is Lipschitz continuous on the set (4.1),

- $u(t)$ does not develop new shocks,

are satisfied on every time interval between two consecutive restartings, thanks to the intermediate strips of width $\delta_2/3$ around the big shocks, and to the choice of $\Delta' t$ in (4.10). After each restarting, we have the situation sketched in Fig. 4.1. Hence in time $\Delta' t$ no small shocks can enter the set (4.2) and no wave, with large negative gradient component u_x^h, can exit from the shock layers $\cup_{i \in \mathcal{S}} \left([y_i - \delta_2, y_i + \delta_2]\right)$. The first lemma provides a lower bound on the time where a gradient catastrophe can occur.

LEMMA 4.2. *Assume* $\bar{t} \in I_{m,h}$, $u(\bar{t}) \in \mathcal{D}^*_{h,L}$. *Then there exists a positive time* $\tau(L)$ *such that* $u(t) \in \mathcal{D}_{h,L}$ *for every* $t \in [\bar{t}, \ \bar{t} + \tau(L)] \cap I_{m,h}$.

Using a comparison argument, as $\tau(L)$ one can take the blow-up time of the solution to the O.D.E. (2.17), with coefficients a, b, c depending only on L.

The next lemma states that, by a suitable restarting procedure, one can approximate any function $u \in \mathcal{D}_{h,L}$ with another function, having almost the same weighted Lipschitz constant, whose gradient satisfies (4.5). The idea is to insert several small h-shocks in regions where the gradient component u_x^h is large and negative.

LEMMA 4.3. *For every $u \in \mathcal{D}_{h,L}$ and $\varepsilon_0 > 0$, there exists $\tilde{u} \in \mathcal{D}^*_{h,L+\varepsilon_0}$ such that $\|\tilde{u} - u\|_{\mathbf{L}^1} < \varepsilon_0$.*

If initially $u(\tau_{m,h-1}) \in \mathcal{D}^*_{h,L} \cap \mathcal{D}'_{h,L}$, we construct an approximate solution on the first subinterval $[\tau_{m,h-1},\ \tau_{m,h-1} + \Delta' t]$ as follows. We choose a time step $\Delta'' t < \tau(L+1)$, say

$$\Delta'' t \doteq \frac{\Delta' t}{N}, \qquad N \doteq 1 + \left[\!\left[\frac{\Delta'' t}{\tau(L+1)} \right]\!\right]. \tag{4.11}$$

We then apply the restarting described in Lemma 4.3 at the times

$$\tau_{m,h-1} + \Delta'' t,\ \tau_{m,h-1} + 2\Delta'' t,\ \dots\ ,\ \tau_{m,h-1} + N\Delta'' t = \tau_{m,h-1} + \Delta t,$$

choosing

$$\varepsilon_0 \doteq \min \left\{ \frac{1}{N},\ \frac{\delta \cdot \Delta'' t}{3} \right\}. \tag{4.12}$$

At the time $t = \tau_{m,h-1} + \Delta' t$, an additional restarting procedure is used.

LEMMA 4.4. *For every $u \in \mathcal{D}_{h,L}$, $\varepsilon_0 > 0$ and every shock y_α, there exists $\tilde{u} \in \mathcal{D}_{h,L+\varepsilon_0}$ and $\delta_3 > 0$ such that $\tilde{u} \equiv u$ on $\mathbb{R} \setminus [y_\alpha - \varepsilon_0,\ y_\alpha + \varepsilon_0]$, while $\tilde{u}$ is constant on the intervals $[y_\alpha - \delta_3,\ y_\alpha[$ and $]y_\alpha,\ y_\alpha + \delta_3]$.*

LEMMA 4.5. *For every $u \in \mathcal{D}_{h,L}$ and $\varepsilon_0 > 0$, there exist $L' > 0$ and $\tilde{u} \in \mathcal{D}^*_{h,L'} \cap \mathcal{D}'_{h,L'}$ such that $\|u - \tilde{u}\|_{L^1} \leq \varepsilon_0$.*

Applying Lemma 4.5 with $\varepsilon_0 = \delta \cdot \Delta' t/3$, the condition (4.3) will hold for all $T \leq \tau_{m,h-1} + \Delta' t$. The solution u is then prolonged to the next subinterval

$$[\tau_{m,h-1} + \Delta' t,\ \tau_{m,h-1} + 2\Delta' t],$$

applying the restarting described in Lemma 4.3 at time steps of some length $\Delta'' t$ (depending on the new Lipschitz constant L'), then the restarting in Lemma 4.5 at the time $\tau_{m,h-1} + 2\Delta' t$, etc... In a finite number of steps, we thus define the approximate solution u on the whole interval $I_{m,h} = [\tau_{m,h-1},\ \tau_{m,h}]$. At the terminal time $\tau_{m,h}$ a third type of restarting is needed.

LEMMA 4.6. *For every $u \in \mathcal{D}_{h,L}$, $\varepsilon_0 > 0$, there exist $L' > 0$ and $\tilde{u} \in \mathcal{D}^*_{h+1,L'} \cap \mathcal{D}'_{h+1,L'}$ such that $\|\tilde{u} - u\|_{L^1} \leq \varepsilon_0$ [if $h = n$ then $\tilde{u} \in \mathcal{D}^*_{1,L'} \cap \mathcal{D}'_{1,L'}$].*

Applying Lemma 4.6 with $\varepsilon_0 = \delta \cdot \delta_1/3n$, the condition (4.3) will hold for all $T \leq \tau_{m,h}$. By induction, this achieves the construction of a δ-accurate approximate ε-solution u for all times $t \in [0, \infty[$.

At this stage, we could consider a sequence of approximate solutions and obtain in the limit an ε-solution of the Cauchy problem (1.1)-(1.2), by a standard compactness argument. Our main goal, however, is to prove the continuous dependence of solutions on their initial data. For this purpose, in the remainder of this section we consider any two δ-accurate approximate ε-solutions u, u' constructed as above (with the same choice of δ_1, δ_2), and give an estimate on the distance $\|u(t,\cdot) - u'(t,\cdot)\|_{\mathbf{L}^1}$ for all $t \geq 0$.

The basic strategy is the following. We consider a Piecewise Regular Path $\gamma_0 \colon [0,1] \to \mathcal{D}^{\delta_2}$ joining $u(0)$ with $u'(0)$. For each $t > 0$ we construct a path $\gamma_t \colon [0,1] \to \mathcal{D}^{\delta_2}$ joining $u(t)$ with $u'(t)$, whose weighted length satisfies

$$\|\gamma_t\|_\star \leq \|\gamma_0\|_\star + C_5 \delta t. \tag{4.13}$$

Recalling definition (1.48), this will provide an estimate on the weighted distance $d_\star\big(u(t), u'(t)\big)$ in terms of the initial distance $d_\star\big(u(0),\, u'(0)\big)$. For $\theta \in [0,1]$, the maps $t \mapsto u^\theta(t) = \gamma_t(\theta)$ are approximate ε-solutions of (1.1), obtained by successive restartings at times $t_1 < t_2 < \cdots$ (the same for all values of θ). By the analysis in Section 3, outside the restarting times the weighted norms of tangent vectors do not increase. Hence the same is true of the weighted length of the path γ_t. In particular, on the interval $]t_{\ell-1},\ t_\ell[$ between any two consecutive restarting times we have

$$\big\|\gamma_{t'}\big\|_\star \leq \big\|\gamma_t\big\|_\star \qquad t_{\ell-1} < t < t' < t_\ell. \tag{4.14}$$

The following analysis will show that the restarting procedures described in Lemmas 4.3–4.6 can be performed simultaneously for all solutions u^θ, in such a way that the length of the path $\gamma_t \colon \theta \mapsto u^\theta(t)$ changes very little across each restarting time. In this way, our paths γ_t will satisfy (4.13) for every $t \geq 0$.

The construction of the paths γ_t is achieved by induction on the intervals $I_{m,h}$, defined at (2.9). Fix an interval $I_{m,h}$. For notational convenience, call $t_* \doteq \tau_{m,h-1}$, $t^* \doteq \tau_{m,h}$. At the initial time t_*, let a piecewise regular path be given: $\gamma_{t_*} : [0,1] \to \mathcal{D}^*_{h,L} \cap \mathcal{D}'_{h,L}$, with $\gamma_{t_*}(0) = u(t_*)$, $\gamma_{t_*}(1) = u'(t_*)$, satisfying (4.13) for $t = t_*$. For every $t \in I_{m,h}$ we will construct a path $\gamma_t \colon [0,1] \to \mathcal{D}^{\delta_2}$ with the following properties:

$$\gamma_t(0) = u(t), \qquad \gamma_t(1) = u'(t), \tag{4.15}$$

$$\gamma_t(\theta) = u^\theta(t) \in \mathcal{D}_{h,L'} \qquad \theta \in [0,1], \tag{4.16}$$

for some Lipschitz constant L'. Moreover, at the final time $t^* = \tau_{m,h}$ we will have

$$\gamma_{t^*}(\theta) = u^\theta(t^*) \in \mathcal{D}^*_{h+1,L''} \cap \mathcal{D}'_{h+1,L''} \tag{4.17}$$

for some constant L'', and the bound (4.13) will hold with $t = t^*$.

Let $\Delta't$ be as in (4.10) and let $\tau(L+1)$ be as in Lemma 4.2. By solving the quasilinear system (3.8) together with the boundary conditions (2.3)-(2.4) and (3.3), (3.9), for each $\theta \in [0,1]$ the corresponding piecewise Lipschitz solution u^θ can be constructed up to the time $t' = \min\big\{t_* + \Delta't,\ t_* + \tau(L+1)\big\}$. By (3.71), the weighted length of the path $\gamma_t \colon \theta \mapsto u^\theta(t)$ does not increase in time. If $\tau(L+1) < \Delta't$ then we apply a first restarting procedure (Restarting 1 below), replacing the old path $\gamma_{t'-}$ with a new path $\gamma_{t'} : [0,1] \to \mathcal{D}^*_{h,L+\varepsilon_0}$ such that

$$\|\gamma_{t'}\|_\star \leq \|\gamma_{t'-}\|_\star + \varepsilon_0, \tag{4.18}$$

$$\|\gamma_{t'}(\theta) - \gamma_{t'-}(\theta)\|_{L^1} < \varepsilon_0 \qquad \theta \in [0,1]. \tag{4.19}$$

We then join $\gamma_{t'}(0)$ with $u(t')$ and $\gamma_{t'}(1)$ with $u'(t')$, by means of two curves whose length is bounded by $C\delta(t'-t_*)$ for some constant $C > 0$. As explained below, for a technical reason the restarting procedure will be applied not exactly at the time $\bar{t} + \tau(L+1)$, but at a suitable time $t' \in [\bar{t} + \tau(L+1)/2,\ \bar{t} + \tau(L+1)]$.
In addition, a similar restarting procedure is performed at each time τ_ℓ where either u or u' is discontinuous, as a function with values in $\mathbf{L}^1$ (i.e., at each time τ_ℓ where a restarting occurred in the original construction of u or u'). In this way, the paths γ_t will be defined for all $t \in [t_*,\ t_* + \Delta' t]$, and satisfy (4.13).

At the time $t' = t_* + \Delta' t$, a different restarting procedure (Restarting 2 below) must be used. This produces a new curve $\gamma_{t'} : [0,1] \to \mathcal{D}^*_{h,L'} \cap \mathcal{D}'_{h,L'}$, for some $L' > 0$, such that (4.18) and (4.19) hold with $\varepsilon_0 > 0$ suitably small. In addition, we construct two small curves joining $\gamma_{t'}(0)$ with $u(t')$ and $\gamma_{t'}(1)$ with $u'(t')$.

The same construction is then repeated on the intervals $[t_* + \Delta' t,\ t_* + 2\Delta' t]$, etc... In a finite number of steps, we thus obtain a path of solutions defined on the whole interval $I_{m,h}$. At the terminal time $t^* \doteq \tau_{m,h}$ a third restarting procedure is used (Restarting 3 below). An appropriate choice of the values of ε_0 at the various restarting times will guarantee the validity of (4.13).

We now describe in detail the three restarting procedures.

Restarting 1

Assume that $\bar{t} \in I_{m,h}$, $\gamma_{\bar{t}}(\theta) \in \mathcal{D}^*_{h,\bar{L}}$ for some $\bar{L} > 0$, and let the restarting procedure occur at some time $t' \in [\bar{t} + \tau(L)/2,\ \bar{t} + \tau(L)]$, for a given $L > \bar{L} > 0$. Moreover, assume that $\gamma_t(\theta) \in \mathcal{D}_{h,L}$ and

$$u^\theta(t,x) = u^{\theta'}(t,x) \qquad x \notin [-M_0,\ M_0],\ \theta, \theta' \in [0,1],$$

for every $t \in [\bar{t} + \tau(L)/2,\ \bar{t} + \tau(L)]$. As in Section 3, we denote by $y_i^\theta(t)$, $i \in \mathcal{S}$, the locations of the big shocks in $u^\theta(t)$, and by $y_\alpha^\theta(t)$, $\alpha \in \mathcal{S}'$, the locations of the small h-shocks. Call $\sigma_i^\theta(t)$, $\sigma_\alpha^\theta(t)$, the size of the shocks located at $y_i^\theta(t)$, $y_\alpha^\theta(t)$, respectively.

REMARK 4.7. Since $u^\theta(t) \in \mathcal{D}$ for all $\theta \in [0,1]$, $t \geq 0$, all these functions have exactly one large i-shock, for each family $i \in \mathcal{S}$. On the other hand, the number of small shocks may vary with θ, t. For notational simplicity, we still write $\mathcal{S}'$ in place of $\mathcal{S}'(\theta,t)$, omitting the explicit dependence on θ, t.

By induction, we assume that the path $\theta \mapsto \gamma_{\bar{t}}(\theta)$ is piecewise regular, i. e. there exist finitely many values $0 = \theta_0 < \theta_1 < \cdots < \theta_r = 1$ such that the restriction of $\gamma_{\bar{t}}$ to each subinterval $]\theta_{j-1},\ \theta_j[$ is a regular path. In particular, the generalized gradient $(v^\theta(\bar{t}), \xi^\theta(\bar{t})) \doteq du^\theta(\bar{t})/d\theta$ is well defined and continuous for $\theta \notin \Theta \doteq \{\theta_0, \theta_1, \ldots, \theta_r\}$.

DEFINITION 4.8. We say that t is an interaction time for the approximate solution u^θ if at time t either two small h-shocks interact, or a small h-shock hits a shock layer or the set (4.6), so that $|y_\alpha^\theta(t) - y_i^\theta(t)| = \delta_2$ or $2\delta_2/3$, for some $\alpha \in \mathcal{S}'$, $i \in \mathcal{S}$.

We define the set

$$\tilde{\Theta} \doteq \{(t,\theta);\ \text{either } \theta \in \Theta \text{ or } t \text{ is an interaction time for } u^\theta\}. \tag{4.20}$$

LEMMA 4.9. *If $(t^\dagger,\theta^\dagger) \notin \tilde{\Theta}$, then the generalized tangent vector*

$$\left(v^\theta(t),\xi^\theta(t)\right) \doteq \frac{du^\theta(t)}{d\theta} \tag{4.21}$$

is well defined and continuous for all (t,θ) in a neighborhood of $(t^\dagger,\theta^\dagger)$.

Define the set

$$J(\varepsilon_0) \doteq \left\{\theta \in [0,1];\ \theta \in \bigcup_{\ell=1}^{r}]\theta_\ell - \varepsilon_0,\ \theta_\ell + \varepsilon_0[\ \right\}. \tag{4.22}$$

Let

$$\bigcup_k [a_k,b_k] \doteq [0,1] \setminus J(\varepsilon_0).$$

By the continuity of the shock strengths, there exists $\bar{\sigma}(\varepsilon_0) > 0$ such that, on every $[a_k,b_k]$, for $\alpha \in \mathcal{S}'$ we have $\left|\sigma^\theta_\alpha(t)\right| > \bar{\sigma}(\varepsilon_0)$.

REMARK 4.10. If t is an interaction time for u^θ, then it may happen that $\mathcal{S}'(t,\theta)$ is not constant near θ. Indeed the number of small shocks may vary along the curve. However, if two small shocks y_α, y_β, interact generating a single shock y', we can avoid changing the set $\mathcal{S}'$ defining $y_\alpha = y_\beta = y'$ after time t. With this notation the set $\mathcal{S}'$ is constant and we obtain the following Lemmas.

LEMMA 4.11. *Let $\xi^\theta_\alpha(t)$, $\xi^\theta_i(t)$, be the shifts of the shocks located at $y^\theta_\alpha(t)$, $y^\theta_i(t)$, respectively. The maps $(t,\theta) \mapsto y^\theta_\alpha(t)$, $y^\theta_i(t)$, are continuous for $t \in [0,1] \setminus J(\varepsilon_0)$. The maps $(t,\theta) \mapsto \sigma^\theta_\alpha(t)$, $\xi^\theta_\alpha(t)$, $\sigma^\theta_i(t)$, $\xi^\theta_i(t)$ are continuous at every point $(t,\theta) \notin \tilde{\Theta}$.*

LEMMA 4.12. *For almost every time t, the set*

$$B(t) \doteq \{\theta;\ (t,\theta) \in \tilde{\Theta}\}$$

is finite.

Using Lemma 4.12, we choose a restarting time $t' \in \left[\bar{t} + \tau(L)/2, \bar{t} + \tau(L)\right]$ such that $B(t')$ is finite. For notational convenience, in the following we use γ to indicate $\gamma(t'-)$ and $\tilde{\gamma}$ to indicate the new path $\gamma(t'+)$ which is produced by the restarting procedure. The next lemma states the continuity of the weight functions W^u_i defined at (3.43).

LEMMA 4.13. *For every $i \in \mathcal{S}$ and $\alpha \in \mathcal{S}'$, the maps $\theta \to W^{\gamma(\theta)}_i(y^\theta_i)$, $\theta \to W^{\gamma(\theta)}_h(y^\theta_\alpha)$, are continuous on $[0,1] \setminus B(t')$.*
For each $j \in \{1,\dots,n\}$, the map $(\theta,x) \mapsto W^{\gamma(\theta)}_j(x)$ is continuous for $\theta \notin B(t')$ and $x \notin \{y^\theta_i,\ y^\theta_i \pm \delta_2,\ y^\theta_\alpha;\ i \in \mathcal{S}, \alpha \in \mathcal{S}'\}$. The map $\theta \mapsto W^{\gamma(\theta)}_j$ is continuous from $[0,1] \setminus B(t')$ into $\mathbf{L}^1_{loc}$.

LEMMA 4.14. *Call (v^θ,ξ^θ) the generalized tangent vector to the map $\theta \mapsto \gamma(\theta)$. Then $v^\theta \in \mathbf{L}^\infty$ for every $\theta \notin B(t')$.*

Consider the set of parameter values

$$\mathcal{K} = \mathcal{K}(\varepsilon_0) \doteq \Big\{]\theta - \varepsilon_0, \theta + \varepsilon_0[\ ; \ \theta \in B(t') \Big\} \subset [0,1].$$

Choose $\delta_3 < \varepsilon_0 \delta_2$ in such a way that the sets $[y_\alpha^\theta - \delta_3, y_\alpha^\theta + \delta_3]$, $\alpha \in \mathcal{S}$, $\theta \notin \mathcal{K}$, are disjoint and do not intersect any of the intervals $[y_i - \delta_2/3,\ y_i + \delta_2/3]$ around big shocks.

LEMMA 4.15. *For every $\varepsilon_0 > 0$, one can partition the interval $[0,1]$ inserting points $0 = \theta_0 < \cdots < \theta_N = 1$ so that*

(i) *For every $\theta \in B(t')$, there exists an index k such that $\theta - \varepsilon_0 = \theta_k$, $\theta + \varepsilon_0 = \theta_{k+1}$.*

In addition, for every $k \in \{1, \ldots, N\}$ such that $[\theta_k,\ \theta_{k+1}] \cap B(t') = \emptyset$, the above partition will satisfy the conditions:

(ii) *For each $i \in \mathcal{S}$ and $\alpha \in \mathcal{S}'$ one has*

$$\sup_{\theta \in [\theta_k, \theta_{k+1}]} |y_i^\theta - y_i^{\theta_k}| < \varepsilon_0 \delta_3, \qquad \sup_{\theta \in [\theta_k, \theta_{k+1}]} |y_\alpha^\theta - y_\alpha^{\theta_k}| < \varepsilon_0 \delta_3. \tag{4.23}$$

(iii) *For every $j = 1, \ldots, n$, the j-th component of $u_x^\theta = \big(\gamma(\theta)\big)_x$ satisfies*

$$\sup_{\theta \in [\theta_k, \theta_{k+1}]} \Big\| \big(\gamma(\theta)\big)_x^j - \big(\gamma(\theta_k)\big)_x^j \Big\|_{L^1} < \varepsilon_0. \tag{4.24}$$

(iv) *For every $x \notin J_k \doteq \{y_i^\theta,\ y_\alpha^\theta\ ;\ i \in \mathcal{S},\ \alpha \in \mathcal{S}',\ \theta \in [\theta_k, \theta_{k+1}]\}$, one has*

$$\sup_{\theta \in [\theta_k, \theta_{k+1}]} \big|\gamma(\theta)(x) - \gamma(\theta_k)(x)\big| < \varepsilon_0 \delta_3. \tag{4.25}$$

(v) *For every $j = 1, \ldots, n$, and every $x \notin J_k^* \doteq \{y_i^\theta,\ y_i^\theta \pm \delta_2,\ y_\alpha^\theta\ ;\ i \in \mathcal{S},\ \alpha \in \mathcal{S}',\ \theta \in [\theta_k, \theta_{k+1}]\}$, one has*

$$\sup_{\theta \in [\theta_k, \theta_{k+1}]} \Big| W_j^{\gamma(\theta)}(x) - W_j^{\gamma(\theta_k)}(x) \Big| < \varepsilon_0. \tag{4.26}$$

Moreover, for every $i \in \mathcal{S}$ and $\alpha \in \mathcal{S}'$ one has

$$\begin{aligned} &\sup_{\theta \in [\theta_k, \theta_{k+1}]} \Big| W_j^{\gamma(\theta)}(y_i^\theta) - W_j^{\gamma(\theta_k)}(y_i^{\theta_k}) \Big| < \varepsilon_0, \\ &\sup_{\theta \in [\theta_k, \theta_{k+1}]} \Big| W_j^{\gamma(\theta)}(y_\alpha^\theta) - W_j^{\gamma(\theta_k)}(y_\alpha^{\theta_k}) \Big| < \varepsilon_0. \end{aligned} \tag{4.27}$$

(vi) *Defining J_k as in (iv), the continuous part v^θ of the generalized tangent vector satisfies*

$$\int_{\theta_k}^{\theta_{k+1}} \int_{\mathbb{R} \setminus J_k} \left| v^\theta(x) - \frac{\gamma(\theta_{k+1})(x) - \gamma(\theta_k)(x)}{\theta_{k+1} - \theta_k} \right| dx\, d\theta < \varepsilon_0 (\theta_{k+1} - \theta_k). \tag{4.28}$$

(vii) *Define $c_k^\theta \doteq (\theta_{k+1} - \theta)/(\theta_{k+1} - \theta_k)$, and let*

$$\tilde{\sigma}_\alpha^\theta \doteq c_k^\theta \sigma_\alpha^{\theta_k} + (1 - c_k^\theta) \sigma_\alpha^{\theta_{k+1}}, \qquad \tilde{\sigma}_i^\theta \doteq c_k^\theta \sigma_i^{\theta_k} + (1 - c_k^\theta) \sigma_i^{\theta_{k+1}},$$

$$\tilde{\xi}_\alpha^\theta \doteq \frac{y_\alpha^{\theta_{k+1}} - y_\alpha^{\theta_k}}{\theta_{k+1} - \theta_k}, \qquad \tilde{\xi}_i^\theta \doteq \frac{y_i^{\theta_{k+1}} - y_i^{\theta_k}}{\theta_{k+1} - \theta_k}.$$

Then for each $i \in \mathcal{S}$, $\alpha \in \mathcal{S}'$ one has

$$|\tilde{\sigma}_i^\theta - \sigma_i^\theta| < \varepsilon_0, \qquad |\tilde{\sigma}_\alpha^\theta - \sigma_\alpha^\theta| < \varepsilon_0, \tag{4.29}$$

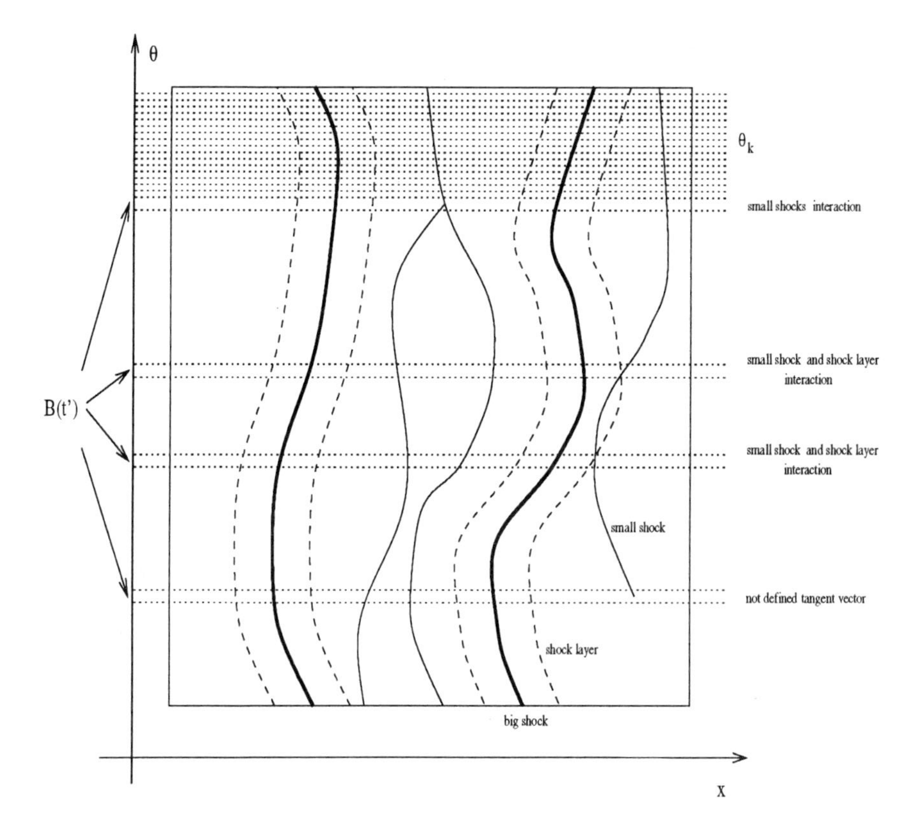

FIGURE 4.2

$$
(4.30)\qquad
\begin{aligned}
&\left|\int_{\theta_k}^{\theta_{k+1}} \xi_\alpha^\theta \sigma_\alpha^\theta W_h^{\gamma(\theta)}(y_\alpha^\theta)d\theta - \int_{\theta_k}^{\theta_{k+1}} \tilde\xi_\alpha^\theta \tilde\sigma_\alpha^\theta W_h^{\gamma(\theta_k)}(y_\alpha^{\theta_k})d\theta\right| < \varepsilon_0(\theta_{k+1}-\theta_k),\\
&\left|\int_{\theta_k}^{\theta_{k+1}} \xi_i^\theta \sigma_i^\theta W_i^{\gamma(\theta)}(y_i^\theta)d\theta - \int_{\theta_k}^{\theta_{k+1}} \tilde\xi_i^\theta \tilde\sigma_i^\theta W_i^{\gamma(\theta_k)}(y_i^{\theta_k})d\theta\right| < \varepsilon_0(\theta_{k+1}-\theta_k).
\end{aligned}
$$

From now on, we consider a fixed partition $0 = \theta_0 < \theta_1 < \cdots < \theta_N = 1$ of $[0,1]$, such that the conclusions of Lemma 4.15 hold (Fig. 4.2). Moreover, we define the set of indices

$$
(4.31)\qquad K \doteq \big\{k;\ [\theta_k, \theta_{k+1}] \cap B(t') \neq \emptyset\big\}.
$$

The path $\tilde\gamma$ will be constructed separately on each interval $[\theta_k, \theta_{k+1}]$. The first step is to apply the restarting procedure described in Lemma 4.3 to each function u^{θ_k}, with a suitable choice of $\delta_4 > 0$. On the "good" intervals $[\theta_k, \theta_{k+1}]$ with $k \notin K$, where the tangent vector $\theta \mapsto (v^\theta, \xi^\theta)$ is continuous, the path $\tilde\gamma$ is defined by suitably interpolating between $\tilde\gamma(\theta_k)$ and $\tilde\gamma(\theta_{k+1})$. A different procedure is used on the "bad" intervals, which intersect $B(t')$. Given $\delta_4 > 0$, let P_{δ_4} be the restarting operator, which associates to every $u \in \mathcal{D}_{h,L}$ a new function $\tilde u \in \mathcal{D}^*_{h,L+\varepsilon_0}$, according to the construction in the proof of Lemma 4.3. If δ_4 is sufficiently small, then the operator P_{δ_4} has a number of nice properties, listed in the following Lemma.

LEMMA 4.16. *For every $\varepsilon_0 > 0$ there exists $\bar{\delta_4}$ such that, for every $\delta_4 \in]0, \bar{\delta_4}]$, the following holds.*

$$\sup_\theta \left| P_{\delta_4}\big(\gamma(\theta)\big)(x) - \gamma(\theta)(x) \right| < \varepsilon_0 \cdot \min_k (\theta_{k+1} - \theta_k) \qquad x \in \mathbb{R}, \tag{4.32}$$

$$\sup_\theta \left\| \big(P_{\delta_4}(\gamma(\theta))\big)_x^j - \big(\gamma(\theta)\big)_x^j \right\|_{\mathbf{L}^1} < \varepsilon_0 \qquad j \neq h, \tag{4.33}$$

$$(V + C_1 Q)\Big(P_{\delta_4}\big(\gamma(\theta)\big)\Big) < (V + C_1 Q)\big(\gamma(\theta)\big) + \varepsilon_0. \tag{4.34}$$

We apply P_{δ_4}, with some $\delta_4 < \bar{\delta}_4$, to every $\gamma(\theta_k)$.

In the next step, we define $\tilde{\gamma}$ on $[\theta_k, \theta_{k+1}]$, $k \notin K$, via suitable interpolations. Since the Rankine-Hugoniot equations are nonlinear, if u^-, u^+ are joined by an h-shock, and v^-, v^+ are also joined by an h-shock, it is not true in general that the convex combinations $\lambda u^- + (1-\lambda)v^-$, $\lambda u^+ + (1-\lambda)v^+$ are joined by an h-shock. For this reason, the interpolation between the values $\tilde{\gamma}(\theta_k)$ and $\tilde{\gamma}(\theta_{k+1})$ must be performed using an alternative coordinate system, where the integral curves of the right eigenvectors r_h coincide with coordinate lines. Moreover, a special construction is needed in a neighborhood of the big shocks y_i, $i \in \mathcal{S}$, and of the small shocks y_α, $\alpha \in \mathcal{S}'$. Care must be taken in order to control the weighted length of the new path, and to ensure that the new functions $\tilde{\gamma}(\theta)$ remain within the domain $\mathcal{D}^{\delta_2}$. Indeed every function $\tilde{\gamma}(\theta)$ must contain a unique large i-shock for every $i \in \mathcal{S}$.

To define a suitable interpolation procedure, consider the set

$$\mathcal{A} \doteq \bigcup_{k \notin K} [\theta_k, \theta_{k+1}] \times \left([-M_0,\ M_0] \setminus \Big(\bigcup_{i \in \mathcal{S}} [y_{k,i}^-, y_{k,i}^+] \ \cup \ \bigcup_{\alpha \in \mathcal{S}'} [y_{k,\alpha}^-, y_{k,\alpha}^+] \Big) \right), \tag{4.35}$$

where

$$\begin{aligned} y_{k,i}^- &\doteq \min\left\{ y_i^{\theta_k}, y_i^{\theta_{k+1}} \right\} - \delta_3, \qquad & y_{k,i}^+ &\doteq \max\left\{ y_i^{\theta_k}, y_i^{\theta_{k+1}} \right\} + \delta_3, \\ y_{k,\alpha}^- &\doteq \min\left\{ y_\alpha^{\theta_k}, y_\alpha^{\theta_{k+1}} \right\} - \delta_3, \qquad & y_{k,\alpha}^+ &\doteq \max\left\{ y_\alpha^{\theta_k}, y_\alpha^{\theta_{k+1}} \right\} + \delta_3. \end{aligned} \tag{4.36}$$

Notice that, since $\delta_3 \leq \varepsilon_0 \delta_2$, for ε_0 sufficiently small the sets $[y_{k,i}^-, y_{k,i}^+]$ are contained inside the set (4.2).

We consider a smooth change of coordinates Φ on Ω such that $D\Phi(u)\big(r_h(u)\big) \equiv \mathbf{e}_1$ for every $u \in \Omega$, where $\mathbf{e}_1 = (1, 0, \ldots, 0)$ is the first unit vector in the canonical basis of $\mathbb{R}^n$. On $\mathcal{A}$ we define

$$\tilde{\gamma}(\theta)(x) \doteq \Phi^{-1}\Big[c_k^\theta \Phi\big(P_{\delta_4}(\gamma(\theta_k))(x)\big) + (1 - c_k^\theta)\Phi\big(P_{\delta_4}(\gamma(\theta_{k+1}))(x)\big) \Big] \tag{4.37}$$

for $\theta \in [\theta_k, \theta_{k+1}]$, where the coefficients c_k^θ were defined in (vii) of Lemma 4.15.

REMARK 4.17. Interpolating as in (4.37), since $y_i^{\theta_k} \neq y_i^{\theta_{k+1}}$ it may happen that, for some $\theta \in [\theta_k, \theta_{k+1}]$, $\tilde{\gamma}(\theta)$ contains a new h-shock inside the set (4.6). However, by (ii) in Lemma 4.15, the distance of this shock from y_i is $\geq 2\delta_2/3 - \varepsilon_0 \Delta t$. Hence, the definition of $\Delta' t$ in (4.10) still ensures that no h-shock can enter the set (4.2).

It remains to define $\tilde{\gamma}$ on the small sets $[y_{k,i}^-, y_{k,i}^+]$, $i \in \mathcal{S}$, and $[y_{k,\alpha}^-, y_{k,\alpha}^+]$, $\alpha \in \mathcal{S}'$. Fix an interval $[\theta_k, \theta_{k+1}]$ and $i \in \mathcal{S}$. Recalling that $c_k^\theta \doteq (\theta_{k+1} - \theta)/(\theta_{k+1} - \theta_k)$,

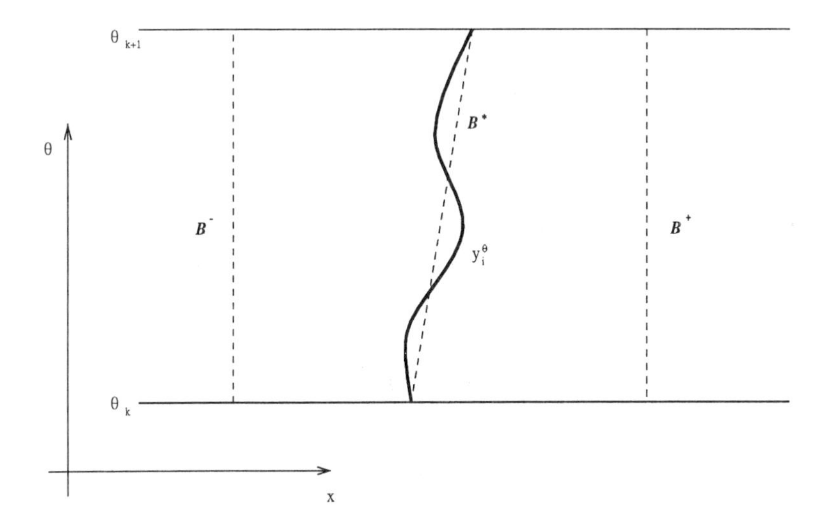

FIGURE 4.3

consider the sets (Fig. 4.3)

$$\mathcal{B} = \mathcal{B}(k,i) \doteq [\theta_k, \theta_{k+1}] \times [y^-_{k,i}, y^+_{k,i}], \tag{4.38}$$

$$\mathcal{B}^* = \mathcal{B}^*(k,i) \doteq \{(\theta, x) \in \mathcal{B};\ x = y^*_{k,i} \doteq c^\theta_k y_i^{\theta_k} + (1 - c^\theta_k) y_i^{\theta_{k+1}}\}, \tag{4.39}$$

$$\begin{aligned} \mathcal{B}^- &= \mathcal{B}^-(k,i) \doteq \{(\theta, x) \in \mathcal{B};\ x = y^-_{k,i}\} \\ \mathcal{B}^+ &= \mathcal{B}^+(k,i) \doteq \{(\theta, x) \in \mathcal{B};\ x = y^+_{k,i}\}. \end{aligned} \tag{4.40}$$

We will define $\tilde{\gamma}$ on $\mathcal{B}$ in such a way that, for every θ, the function $\tilde{\gamma}(\theta)$ has a unique shock on the set $[y^-_{k,i},\ y^+_{k,i}]$, located at $y^*_{k,i}$. Since the values of $\tilde{\gamma}$ have already been defined by (4.37) on $\mathcal{B}^- \cup \mathcal{B}^+$, our goal can be achieved by assigning the right and left limits on $\mathcal{B}^*$ and then interpolating linearly w.r.t. the x-variable. For $(\theta, x) \in \mathcal{B}^*$ we thus define

$$\begin{aligned} \tilde{\gamma}(\theta, x-) &\doteq c^\theta_k \tilde{\gamma}(\theta_k)(y_i^{\theta_k}-) + (1 - c^\theta_k)\tilde{\gamma}(\theta_{k+1})(y_i^{\theta_{k+1}}-), \\ \tilde{\gamma}(\theta, x+) &\doteq \Psi^\varepsilon_i(\tilde{\sigma}^\theta_i)\big(\tilde{\gamma}(\theta)(x-)\big), \end{aligned} \tag{4.41}$$

where Ψ^ε_i was defined at (2.2) and $\tilde{\sigma}^\theta_i$ in (vii) of Lemma 4.15. This definition clearly guarantees that $\tilde{\gamma}(\theta)$ has an i-shock located at the point $y^*_{k,i}$. Next, we linearly interpolate between the values $\tilde{\gamma}(\theta)(y^-_{k,i})$, $\tilde{\gamma}(\theta)(x-)$ over the interval where $x \in [y^-_{i,k},\ y^*_{k,i}[$, and between the values $\tilde{\gamma}(\theta)(x+)$, $\tilde{\gamma}(\theta)(y^+_{k,i})$ over the interval where $x \in]y^*_{k,i},\ y^+_{i,k}]$.

REMARK 4.18. Notice that we slightly modified $\tilde{\gamma}(\theta_k)$. However, to ensure that $\tilde{\gamma}$ is well defined, it is enough to change the data $\tilde{\gamma}(\theta_k)$, by linear interpolation, on the sets $[y^-_{k,i}, y^+_{k,i}]$ before starting the interpolating procedure.

The definition of $\tilde{\gamma}$ in a neighborhood of a small shock y^θ_α, $\alpha \in \mathcal{S}'$ is entirely similar.

Thanks to the above construction, the functions $\tilde{\gamma}(\theta)$ have a number of useful properties.

LEMMA 4.19. *On the set $\mathcal{A}$ there holds*

$$\left|\tilde{\gamma}(\theta)(x) - \gamma(\theta)(x)\right| < C(\varepsilon_0 + \delta_4). \tag{4.42}$$

LEMMA 4.20. *For every $j \neq h$, on the set $\mathcal{A}$ one has*

$$\left\| \left(\tilde{\gamma}(\theta)\right)_x^j - \left(\gamma(\theta)\right)_x^j \right\|_{\mathbf{L}^1} < C(\varepsilon_0 + \delta_4). \tag{4.43}$$

LEMMA 4.21. *For every $k \notin K$, every $\theta \in [\theta_k, \theta_{k+1}]$, $j = 1, \ldots, n$, and every $x \in \mathcal{A}$ except at most a set of measure $< C\varepsilon_0 \Delta t$, one has*

$$\left| W_j^{\gamma(\theta)}(x) - W_j^{\tilde{\gamma}(\theta)}(x) \right| \le C(\varepsilon_0 + \delta_4), \qquad i = 1, \ldots, n. \tag{4.44}$$

LEMMA 4.22. *For some constant $C = C(L)$ one has*

$$\sum_{j=1}^{n} \left| \int_{\mathcal{A}} |v_j(x)| W_j^{\gamma(\theta)}(x) dx d\theta - \int_{\mathcal{A}} |\tilde{v}_j(x)| W_j^{\tilde{\gamma}(\theta)}(x) dx d\theta \right| < C(\varepsilon_0 + \delta_4). \tag{4.45}$$

LEMMA 4.23. *For some constant $C = C(L, \|v\|_{\mathbf{L}^\infty})$ there holds*

$$\max_{j=1,\ldots,n} \sup_{(\theta,x)\in\mathcal{A}} \left| \left(\tilde{\gamma}(\theta)\right)_x^j (x) \exp[\beta W_j^{\tilde{\gamma}(\theta)}(x)] \right| < \bar{L} + C(\varepsilon_0 + \Delta t + \delta_4). \tag{4.46}$$

We now complete the construction of $\tilde{\gamma}$ on the "bad" set $\mathcal{K}$, using the following lemma.

LEMMA 4.24. *For every k such that $]\theta_k, \theta_{k+1}[\subset \mathcal{K}$ there exists a piecewise regular curve γ_k, joining $\tilde{\gamma}(\theta_k)$ with $\tilde{\gamma}(\theta_{k+1})$, such that $\|\gamma_k\|_\star < C\varepsilon_0$, $\|\gamma_k(\theta) - \gamma(\theta_k)\|_{\mathbf{L}^1} < C\varepsilon_0$ and $\gamma_k(\theta) \in \mathcal{D}^*_{h,L+\varepsilon_0}$ for every $\theta \in [\theta_k, \theta_{k+1}]$.*

CLAIM 4.25. *The estimates (4.18) and (4.19) hold for δ_3, δ_4 and ε_0 sufficiently small. Moreover the new path $\tilde{\gamma}$ is piecewise regular.*

CLAIM 4.26. *If ε_0 and δ_4 are small enough then, for every $\theta \in [0, 1]$, one has*

$$\text{w-Lip}(\tilde{\gamma}(\theta)) < \bar{L} + \frac{1}{N}, \tag{4.47}$$

$$(V + C_1 Q)(\tilde{\gamma}(\theta)) < \delta_0, \tag{4.48}$$

where N is defined in (4.11). Moreover, $\tilde{\gamma}(\theta) \in \mathcal{D}^*_{h,L+1/N}$.

Claims 4.25 and 4.26 will be proved at the end of this section.
Now we have to construct the small curves γ_1 and γ_2 connecting $\tilde{\gamma}(0)$ with $u(t')$ and $\tilde{\gamma}(1)$ with $u'(t')$. Let us describe the construction of γ_1, since the construction of γ_2 is entirely similar. If t' is not a restarting time for u, then $\tilde{\gamma}(0)$ is obtained from $u(t')$ applying the operator P_{δ_4} and the small modifications described above. In particular $u(t')$ and $\tilde{\gamma}(0)$ contain the same large shocks at the same locations y_i. We let γ_1 be the path

$$\theta \mapsto \Phi^{-1}\Big(\theta \Phi\big(u(t')\big) + (1-\theta)\Phi\big(\tilde{\gamma}(0)\big)\Big), \qquad \theta \in [0, 1].$$

Since no shock is shifted we easily obtain

$$\|\gamma_1\|_\star < C \big\| u(t') - \tilde{\gamma}(0) \big\|_{L^1}$$

for some constant C. If t' is a restarting time for u, we use a similar construction joining $\tilde{\gamma}(0)$ with $u(t'+)$. In this way we obtain

$$\|\gamma_1\|_\star < C\Big(\big\|u(t'+) - u(t'-)\big\|_{L^1} + \big\|u(t'-) - \tilde{\gamma}(0)\big\|_{L^1}\Big) \leq 2C\delta(t' - t^*).$$

The other relevant properties for γ_1 can be proved as for $\tilde{\gamma}$. Finally, we reparametrize the path obtained by concatenating γ_1, $\tilde{\gamma}$ and γ_2, and obtain a path defined on $[0,1]$.

REMARK 4.27. Notice that, if t' is not a restarting time for u (respectively u'), then $\gamma_1(\theta) \notin \mathcal{D}^*_{h,L+1/N}$ (respectively $\gamma_2(\theta) \notin \mathcal{D}^*_{h,L+1/N}$) for some θ. Indeed, $u(t')$ does not necessarily satisfy the condition

$$u^h_x(t',x)\exp[\beta W_h^{u(t')}(x)] \geq -1.$$

However, since u is an approximate solution, it does not develop new shocks up to a certain time $\tilde{t} > t'$ at which a restarting happens. Let

$$M \doteq \inf u^h_x(t',x)\exp[\beta W_h^{u(t')}(x)],$$

where the inf is taken outside the set (4.6). Since $\tilde{\gamma}(0)$ satisfies the condition (4.5), then for every θ it follows

$$(\gamma_1(\theta))^h_x(x)\exp[\beta W_h^{\gamma_1(\theta)}(x)] \geq M - C\varepsilon_0.$$

From the proof of Lemma 4.2, we have that, for ε_0 small enough, the approximate solution corresponding to $\gamma_1(\theta)$ is well defined up to time $\tilde{t}$, hence up to the next restarting time.

This completes the first restarting procedure.

Restarting 2

We follow the same procedure of Restarting 1 with the following differences. We choose t' such that $B(t')$ is finite and fix $\Delta t < \varepsilon_0\delta_2$. Again, the set $[0,1]$ is partitioned inserting points θ_k in such a way that the conclusions of Lemma 4.15 hold. Now, the operator P_{δ_4} is not applied. While, for every k, we modify $\gamma(\theta_k)$ via the construction of Lemma 4.5 with parameter $\Delta t/2$. On the set $\mathcal{A}$, $\tilde{\gamma}$ is defined linearly interpolating on the set $[\theta_k, \theta_{k+1}]$, $k \notin K$. On the sets $\mathcal{B}(k,i)$ the same procedure is used.
Fix now $k \notin K$ and $\alpha \in \mathcal{S}'$ such that $y^{\theta_k}_\alpha$ lies inside the set (4.6). In the construction of $\tilde{\gamma}$, the shock is replaced by a smooth compressive wave on the set $[y^{\theta_k}_\alpha - \Delta t/2,\ y^{\theta_k}_\alpha + \Delta t/2]$. Recall the definition of c^θ_k. For $\theta \in [\theta_k, \theta_{k+1}]$ and $x \in [-\Delta t/2, \Delta t/2]$, we define:

$$\begin{aligned}\tilde{\gamma}(\theta)\Big(c^\theta_k\ (y^{\theta_k}_\alpha + x) + (1 - c^\theta_k)\ (y^{\theta_{k+1}}_\alpha + x)\Big) &\doteq \\ &\doteq c^\theta_k\ \tilde{\gamma}(\theta_k)(y^{\theta_k}_\alpha + x) + (1 - c^\theta_k)\ \tilde{\gamma}(\theta_{k+1})(y^{\theta_{k+1}}_\alpha + x).\end{aligned}$$

In this way, $\tilde{\gamma}(\theta)$ has a compressive wave on the set

$$[c^\theta_k y^{\theta_k}_\alpha + (1 - c^\theta_k)y^{\theta_{k+1}}_\alpha - \Delta t/2, c^\theta_k y^{\theta_k}_\alpha + (1 - c^\theta_k)y^{\theta_{k+1}}_\alpha + \Delta t/2],$$

that is obtained interpolating the compressive waves of $\tilde{\gamma}(\theta_k)$ and $\tilde{\gamma}(\theta_{k+1})$. We complete the construction on the set $\mathcal{B}(k,\alpha)$ (defined similarly to $\mathcal{B}(k,i)$) via linear interpolation w.r.t. the x variable.

The conclusions of Lemma 4.24 are still valid and the proof is similar. Claim 4.25 and (4.48) can be proved in the same way thanks to the accurate construction on the sets $\mathcal{B}(k,\alpha)$.
The construction of the curves γ_1, γ_2, can be done via linear interpolation. Indeed, since $\Delta' t$ of (4.10) depends only on δ_1, δ_2 (and not on L), t' is a restarting time also for u and u'.
We produced a curve with values in $\mathcal{D}'_{h,L'}$. Now, applying Restarting 1 again, we are done.

Restarting 3

We first construct a new curve with values in $\mathcal{D}_{h+1,L'}$ $[\mathcal{D}_{1,L'}$ if $h=n]$ for some L'. Then we apply again Restarting 1 in order to obtain a curve with values in $\mathcal{D}^*_{h+1,L'+\varepsilon_0}$ $[\mathcal{D}^*_{1,L'+\varepsilon_0}$ if $h=n]$.
For the first step, we follow the same procedure of Restarting 2 with the following differences. We apply the construction of Lemma 4.5 to every shock $y_\alpha^{\theta_k}$. Again, the operator P_{δ_4} is not applied, so that the only discontinuities of each $\tilde{\gamma}(\theta_k)$ are the big shocks. In particular $\tilde{\gamma}(\theta_k) \in \mathcal{D}_{h+1,L'}$ for some $L'>0$.

This completes the analysis of the restarting procedures. The remainder of the section contains proofs of the various lemmas stated above.

Proofs

PROOF OF LEMMA 4.2. Let us define the functions

$$w^i(t,x) \doteq u_x^i(t,x)\exp\left[\beta W_i^{u(t)}(x)\right], \quad t\geq 0,\ x\in\mathbb{R},\ i=1,\dots,n,$$

where the weights W_i^u are defined in (3.43), (3.44).
By assumption we have that $\max_i \sup_x \left|w^i(\bar{t},x)\right| < L$. We will achieve the proof by contradiction.
Assume that there exist $i\in\{1,\dots,n\}$ and a point (t_1,y), $t_1>\bar{t}$, such that

$$\left|w^i(t_1,y)\right| = L, \quad \left|w^j(t,x)\right| < L\ \forall t\in[\bar{t},t_1[, \qquad \forall x\in\mathbb{R},\ \forall j=1,\dots,n. \tag{4.49}$$

Let us consider first the case in which $u(t_1)$ is continuous at y. Let $x=x_i(t)$ be the i–characteristic curve passing through (t_1,y). Let $\tau'\in]\bar{t},t_1[$, suitably close to t_1, such that

$$\left|w^i(t,x_i(t))\right| \geq \frac{1}{2}\left|w^i(t_1,y)\right|, \quad \forall t\in[\tau',t_1], \tag{4.50}$$

and $x_i(t)$ does not intersect any shock for $t\in[\tau',t_1]$. Differentiating the map $t\mapsto w^i(t,x_i(t))$ one obtains

$$\begin{aligned}\frac{d}{dt}w^i(t,x_i(t)) &= \left[\frac{d}{dt}u_x^i + \beta u_x^i\frac{d}{dt}W_i^u\right]\exp(\beta W_i^u) = \\ &= \left[\left(\sum_{j<k}G_{ijk}u_x^ju_x^k - \sum_j(\nabla\lambda_i\cdot r_j)u_x^iu_x^j\right) + \beta u_x^i\frac{d}{dt}W_i^u\right]\exp(\beta W_i^u).\end{aligned} \tag{4.51}$$

Let $\kappa > 0$ be the infimum of the quantities $\left|\lambda_j^{(h)}(u) - \lambda_k^{(h)}(u)\right|$ for $j \neq k$ and $u \in \Omega$. From (3.36) and (3.46), along the characteristic $x = x_i(t)$ one has

$$\begin{aligned} \frac{d}{dt}W_i^u =& \kappa_1 \frac{d}{dt}R_i^u + \kappa_1\kappa_2\frac{d}{dt}Q(u) \leq \\ \leq& \kappa_1\left[-\kappa\sum_{j\neq i}|u_x^j| + C\Lambda(u)\right] - \kappa_1\kappa_2\frac{C_0}{2}\Lambda(u) \leq \\ \leq& -\kappa\kappa_1\sum_{j\neq i}|u_x^j|, \end{aligned} \tag{4.52}$$

provided that $\kappa_2 \geq 2C/C_0$. Let $M \doteq \sup_{i,j,k,|u|\leq\delta_0}|G_{ijk}(u)|$.
If $i \neq h$, from (4.52) and the fact that $\nabla\lambda_i \equiv 0$, one has

$$\begin{aligned} &\frac{d}{dt}|w^i(t,x_i(t))| \\ &\leq \left[M\sum_{j<k}|u_x^j||u_x^k| + C\sum_{j\neq i}|u_x^i||u_x^j| - \beta\kappa\kappa_1\sum_{j\neq i}|u_x^i||u_x^j|\right]\exp(\beta W_i^u). \end{aligned} \tag{4.53}$$

Observe that, by (4.50),

$$\left|w^j(t,x)\right| \leq \left|w^i(t_1,y)\right| \leq 2\left|w^i(t,x_i(t))\right|,$$

for every $x \in \mathbb{R}$, $t \in [\tau', t_1]$, $j = 1, \dots, n$. Henceforth, we can choose β large enough, depending only on κ, κ_1, κ_2 and M in such a way that

$$\frac{d}{dt}\left|w^i(t,x_i(t))\right| \leq 0, \quad \forall t \in [\tau', t_1],$$

in contradiction with (4.49).
In the case $i = h$, if $w^h(t_1,y) = L$ then the inequality (4.53) still holds, since $\nabla\lambda_h \cdot r_h \geq 0$, and the contradiction is reached as above. It remains to analyze the case $w^h(t_1,y) = -L$. Since $u \in \mathcal{D}^*_{h,L}$, we have that $w^h(\bar{t},x) \geq -1$ for every $x \in \mathbb{R}$. It is not restrictive to assume $w^h(t,x_h(t)) < 0$ for every $t \in [\bar{t}, t_1]$. From (4.51) we obtain

$$\begin{aligned} \frac{d}{dt}|w^h(t,x_h(t))| =& -\frac{d}{dt}w^h(t,x_h(t)) = \\ =& \Big[-\sum_{j<k}G_{ijk}u_x^j u_x^k + \sum_{j\neq h}(\nabla\lambda_h \cdot r_j)u_x^h u_x^j \\ &+ (\nabla\lambda_h \cdot r_h)(u_x^h)^2 - \beta u_x^h\frac{d}{dt}W_h^u\Big]\exp(\beta W_h^u). \end{aligned} \tag{4.54}$$

It is possible to define some constants a, b and c, depending only on L and the coefficients of (4.54), such that $\left|w^h(t,x_h(t))\right|$ is bounded from above by the solution of the Cauchy problem

$$\dot{z} = az^2 + bz + c, \qquad z(\bar{t}) = 1,$$

We can now define

$$\tau(L) = \frac{1}{2}\inf\left\{s > 0;\ z(\bar{t} + s) = L\right\}.$$

Clearly $\left|w^h(t,x_h(t))\right| < L$ for every $t \in [\bar{t}, \bar{t} + \tau(L)]$, completing the proof in this case.

Let us examine what happens when a i-characteristic curve crosses a small h-shock of strength σ, say at the point x. Let us call u_x^{j-}, u_x^{j+} respectively the j-th component of u_x to the left and to the right of the shock. In the same manner we define $w^{j\pm}$.
If $\mathcal{I}$ and $\mathcal{O}$ denote respectively the set of incoming and outgoing components, from (3.23) and (3.27) we deduce that

$$u_x^{j\pm} = \sum_{i^\pm \in \mathcal{I}} \frac{\partial U^j}{\partial u_x^{i\pm}} u_x^{i\pm} = \sum_{i>h} \frac{\partial U^j}{\partial u_x^{i-}} u_x^{i-} + \sum_{i<h} \frac{\partial U^j}{\partial u_x^{i+}} u_x^{i+}, \quad j^\pm \in \mathcal{O}.$$

If $|w^{i\pm}| \leq w$ for every $i^\pm \in \mathcal{I}$, one obtains

$$|u_x^{j\pm}| \leq w \left[\sum_{i>h} \left| \frac{\partial U^j}{\partial u_x^{i-}} \right| \mathrm{e}^{-\beta W_i^u(x-)} + \sum_{i<h} \left| \frac{\partial U^j}{\partial u_x^{i+}} \right| \mathrm{e}^{-\beta W_i^u(x+)} \right], \quad j^\pm \in \mathcal{O}.$$

Recalling (3.24) and (3.25), this implies

$$|w^{j\pm}| \leq \big(1 + C'|\sigma|\big) w \exp\big[\beta(W_j^u(x\pm) - W_j^u(x\mp))\big]. \tag{4.55}$$

From (3.43) we have $W_j^u(x\pm) - W_j^u(x\mp) \leq -\kappa_1|\sigma|$, so that, choosing $\beta \geq C'/\kappa_1$, we obtain

$$\big(1 + C'|\sigma|\big) \exp\big[\beta(W_j^u(x\pm) - W_j^u(x\mp))\big] \leq 1. \tag{4.56}$$

Finally, from (4.55) and (4.56), we deduce

$$\max_{j\pm \in \mathcal{O}} |w^{j\pm}| \leq \max_{i\pm \in \mathcal{I}} |w^{i\pm}|. \tag{4.57}$$

If $u(t_1)$ is discontinuous at x, then (4.57) gives the contradiction. □

PROOF OF LEMMA 4.3. Let $y_{j_1} < \cdots < y_{j_\nu}$ be the positions of the large jumps of u, and define $y_0 \doteq -M_0 - \delta_2$, $y_{\nu+1} \doteq M_0 + \delta_2$. Moreover, let $I_i \doteq [y_i + 2\delta_2/3, y_{i+1} - 2\delta_2/3]$, $i \in \mathcal{S}$.

Let $\delta_4 > 0$ be given. For a fixed index i, let $\Delta \doteq \{x_0, \ldots, x_N\}$, $y_i + 2\delta_2/3 = x_0 < \cdots < x_N = y_{i+1} - 2\delta_2/3$ be a partition of I_i such that

$$\big(1 + \text{w-Lip}(u)\big) \cdot \max\{x_{r+1} - x_r;\ r = 0, \ldots, N-1\} < \delta_4$$

and $\{y_\alpha \in I_i;\ \alpha \in \mathcal{S}'\} \subset \Delta$, where the y_α are the locations of the small h-shocks.

On the interval I_i we define $\tilde{u}$ as follows. On every subinterval $[x_r, x_{r+1}[$ let $U(x)$ be the solution of

$$\frac{d}{dx} U(x) = r_h(U(x))[u_x^h(x)]_+ + \sum_{j \neq h} r_j(U(x)) u_x^j(x), \quad U(x_r) = u(x_r). \tag{4.58}$$

Now let us define

$$\tilde{u}(x) \doteq U(x) + \frac{x - x_r}{x_{r+1} - x_r}\big[(\exp q r_h) u(x_{r+1}) - U(x_{r+1})\big], \qquad x \in [x_r, x_{r+1}[,$$

where $q \doteq \int_{x_r}^{x_{r+1}} [u_x^h(x)]_- \, dx$. In this way we can define $\tilde{u}$ on $\cup_i I_i$, and we set $\tilde{u} = u$ outside $\cup_i I_i$.

Let $u_r \doteq u(x_r)$, $U_r \doteq U(x_r)$. We want to estimate the difference $(\exp q r_h) u_{r+1} - U_{r+1}$. By the definition of the exponential map, we have that

$$(\exp q r_h) u_{r+1} = u_{r+1} + q r_h(u_{r+1}) + O(q^2).$$

On the other hand, from (4.58) one has

$$\begin{aligned}U_{r+1} =& u_r + \int_{x_r}^{x_{r+1}} \sum_{j\neq h} r_j(u)u_x^j\,dx + \int_{x_r}^{x_{r+1}} \sum_{j\neq h} \big(r_j(U) - r_j(u)\big)u_x^j\,dx + \\ &+ \int_{x_r}^{x_{r+1}} r_h(u_{r+1})[u_x^h]_-\,dx + \int_{x_r}^{x_{r+1}} \big(r_h(U) - r_h(u_{r+1})\big)[u_x^h]_-\,dx = \\ =& u_{r+1} + qr_h(u_{r+1}) + \int_{x_r}^{x_{r+1}} \sum_{j\neq h} \big(r_j(U) - r_j(u)\big)u_x^j\,dx \\ &+ \int_{x_r}^{x_{r+1}} \big(r_h(U) - r_h(u_{r+1})\big)[u_x^h]_-\,dx.\end{aligned}$$

We thus obtain

$$\begin{aligned}\big|(\exp qr_h)u_{r+1} - U_{r+1}\big| \leq& CL\|U-u\|_{\mathbf{L}^\infty(x_r,x_{r+1})}\cdot(x_{r+1}-x_r) \\ &+ O\big((x_{r+1}-x_r)^2\big).\end{aligned} \tag{4.59}$$

Let us define the absolutely continuous function $z(x) \doteq \big|U(x) - u(x)\big|$, $x \in [x_r, x_{r+1}[$. We have that $z(x_r) = 0$, and

$$\begin{aligned}\frac{d}{dx}z(x) \leq& \big|u_x(x) - U_x(x)\big| \leq \\ \leq& \left|-\big[u_x^h(x)\big]_- r_h\big(u(x)\big) + \sum_j U_x^j(x)\big(r_j(u(x)) - r_j(U(x))\big)\right| \leq \\ \leq& CL + CLz(x).\end{aligned}$$

By Gronwall's inequality one obtains, for every $x \in [x_r, x_{r+1}[$,

$$\big|u(x) - U(x)\big| \leq e^{CL(x-x_r)} - 1 \leq C'L(x_{r+1} - x_r). \tag{4.60}$$

From this last inequality and (4.59) we deduce that, for $x \in [x_r, x_{r+1}[$,

$$\begin{aligned}\big|u(x) - \tilde u(x)\big| \leq& |u(x) - U(x)| + |\tilde u(x) - U(x)| \leq \\ \leq& z(x) + \big|(\exp qr_h)u_{r+1} - U_{r+1}\big| \leq \\ \leq& CL(x_{r+1} - x_r) + O((x_{r+1}-x_r)^2) \leq C\delta_4 + O(\delta_4^2).\end{aligned} \tag{4.61}$$

It is clear that this inequality holds for every $x \in \cup_i I_i$. Since $u = \tilde u$ outside $\cup_i I_i$, the inequality $\|u - \tilde u\|_{\mathbf{L}^1} < \varepsilon_0$ follows by choosing δ_4 small enough.

Combining (4.59) and (4.60), we obtain

$$\big|(\exp qr_h)u_{r+1} - U_{r+1}\big| \leq C\delta_4(x_{r+1} - x_r). \tag{4.62}$$

Since, for $x \in [x_r, x_{r+1}[$,

$$\tilde u_x(x) = U_x(x) + \frac{1}{x_{r+1} - x_r}\big((\exp qr_h)u_{r+1} - U_{r+1}\big), \tag{4.63}$$

we deduce that

$$|\tilde u_x| \leq |U_x| + C\delta_4 + O(\delta_4^2) \leq |u_x| + C\delta_4 + O(\delta_4^2). \tag{4.64}$$

Hence, since $U_x^h = 0$

$$\begin{aligned}\tilde u_x^h \exp[\beta W_h^{\tilde u}] \geq& -C\Big(|\tilde u_x^h - U_x^h|\Big) \geq \\ \geq& -C\Big(|\tilde u - U| + |\tilde u_x - U_x|\Big) \geq -C'\delta_4,\end{aligned}$$

for some $C' > 0$.
To prove that $\tilde{u} \in \mathcal{D}^*_{h,L+\varepsilon_0}$, it remains to check the value of the Glimm functional and of the weighted Lipschitz constant. We claim that $(V + C_1Q)(\tilde{u}) < \delta_0$. Indeed $V(\tilde{u}) \le V(u) + C\delta_4$. Moreover, the effect of the transformation $u \mapsto \tilde{u}$ is to shift the negative h-wave in a small interval (of measure $\le \delta_4$) into a small h-shock. Hence $Q(\tilde{u}) \le Q(u) + CL\delta_4$. If δ_4 is small enough we thus obtain $(V + C_1Q)(\tilde{u}) \le (V + C_1Q)(u) + C'\delta_4 < \delta_0$. In the same way we obtain $\left|W_i^u(x) - W_i^{\tilde{u}}(x)\right| \le C\delta_4$, and hence, if δ_4 is small enough, we have w-Lip$(\tilde{u}) \le L + \varepsilon_0$. □

PROOF OF LEMMA 4.4. Choose $\delta_3 > 0$, and define

$$u_\alpha(x) \doteq \begin{cases} u\left(\frac{x - y_\alpha\sqrt{\delta_3} + \delta_3}{1-\sqrt{\delta_3}}\right), & \text{if } x \in]y_\alpha - \sqrt{\delta_3}, y_\alpha - \delta_3[, \\ u\left(\frac{x - y_\alpha\sqrt{\delta_3} - \delta_3}{1-\sqrt{\delta_3}}\right), & \text{if } x \in]y_\alpha + \delta_3, y_\alpha + \sqrt{\delta_3}[, \\ u(y_\alpha-), & \text{if } x \in [y_\alpha - \delta_3, y_\alpha[, \\ u(y_\alpha+), & \text{if } x \in]y_\alpha, y_\alpha + \delta_3]. \end{cases}$$

Define the function

$$\tilde{u}(x) \doteq \begin{cases} u_\alpha, & \text{if } x \in I_\alpha, \\ u(x), & \text{otherwise.} \end{cases}$$

We prove that $(V + C_1Q)(\tilde{u}) < \delta_0$. Reasoning as in Lemma 4.3, it can be shown that $V(\tilde{u}) \le V(u) + C\delta_3$, $Q(\tilde{u}) \le Q(u) + C\delta_3$, so that it suffices to choose δ_3 small enough. Finally

$$\text{w-Lip}(\tilde{u}) \le \frac{\text{w-Lip}(u)}{1 - \sqrt{\delta_3}} + C\delta_3 \le L + \varepsilon_0$$

for δ_3 sufficiently small. □

PROOF OF LEMMA 4.5. Choose $\delta_3 > 0$, and for every $\alpha \in \mathcal{S}'$ such that y_α belongs to the set (4.6), define

$$u_\alpha(x) \doteq \begin{cases} u\left(\frac{x - y_\alpha\sqrt{\delta_3} + \delta_3}{1-\sqrt{\delta_3}}\right), & \text{if } x \in]y_\alpha - \sqrt{\delta_3}, y_\alpha - \delta_3[, \\ u\left(\frac{x - y_\alpha\sqrt{\delta_3} - \delta_3}{1-\sqrt{\delta_3}}\right), & \text{if } x \in]y_\alpha + \delta_3, y_\alpha + \sqrt{\delta_3}[, \\ (\exp(-\sigma_\alpha(x)r_h))u(y_\alpha-), & \text{if } x \in [y_\alpha - \delta_3, y_\alpha + \delta_3], \end{cases} \tag{4.65}$$

where $\sigma_\alpha(x) \doteq \sigma_\alpha(x - y_\alpha + \delta_3)/(2\delta_3)$.
Let δ_3 be small enough such that the sets $I_\alpha \doteq \left[y_\alpha - \sqrt{\delta_3}, y_\alpha + \sqrt{\delta_3}\right]$, $\alpha \in \mathcal{S}_h$, are pairwise disjoint and do not intersect the set (4.2). Let us define the function

$$u'(x) \doteq \begin{cases} u_\alpha, & \text{if } x \in I_\alpha \text{ for some } \alpha \in \mathcal{S}_h, \\ u(x), & \text{otherwise.} \end{cases}$$

Clearly $u' \in \mathcal{D}'_{h,L'}$ for some $L' > 0$. Moreover $\|\tilde{u} - u\|_{\mathbf{L}^1} \le C\delta_3$ for some $C > 0$. We now apply to u' the construction of Lemma 4.3 restricted to the complement of the set (4.6), obtaining $\tilde{u}$. The estimate $(V + C_1Q)(\tilde{u}) < \delta_0$ is obtained as in Lemmas 4.3 and 4.4. □

PROOF OF LEMMA 4.6. Choose $\delta_3 > 0$, and for every $\alpha \in \mathcal{S}_h$ define u_α as in (4.65). Let

$$u'(x) \doteq \begin{cases} u_\alpha, & \text{if } x \in I_\alpha \text{ for some } \alpha \in \mathcal{S}', \\ u(x), & \text{otherwise.} \end{cases}$$

Now u' is discontinuous only at the big shocks y_i, $i \in \mathcal{S}$, hence $u' \in \mathcal{D}_{h+1,L'}$ for some $L' > 0$. We apply the construction of Lemma 4.3 for the $(h+1)$-th family obtaining the required function $\tilde{u}$. The estimates on $\|\tilde{u} - u\|_{\mathbf{L}^1}$, on w-Lip$(\tilde{u})$ and on $(V + C_1 Q)(\tilde{u})$ follows from the corresponding estimates in Lemmas 4.3 and 4.4. □

PROOF OF LEMMA 4.9. Since $(t^\dagger, \theta^\dagger) \notin \tilde{\Theta}$ then the tangent vector is well defined. We have that $(v^\theta(\bar{t}), \xi^\theta(\bar{t}))$ is continuous at $\theta^\dagger$ as a function of θ and it satisfies equations (1.34), (1.35) and (1.36). Moreover the interactions between small shocks do not produce new waves. Hence the conclusion follows from the analysis in [**B-M1**]. □

PROOF OF LEMMA 4.11. Since $\sigma_\alpha^\theta > \bar{\sigma}(\varepsilon_0) > 0$, the function y_α^θ must be continuous in θ. Indeed, assume that there exists a sequence θ_μ, $\theta_\mu \to \tilde{\theta}$, such that $y_\alpha^{\theta_\mu} \to \bar{y} \neq y_\alpha^{\tilde{\theta}}$. There exists a constant $C > 0$ such that $\big|\gamma(\theta)(y_\alpha^\theta -) - \gamma(\theta)(y_\alpha^\theta +)\big| > C\sigma(\varepsilon_0)$. Since $\gamma(\theta)$ is continuous in $\mathbf{L}^1$, $\gamma(\theta_\mu)(x) \to \gamma(\tilde{\theta})(x)$ for almost every x. There exist $x^- < \bar{y} < x^+$, $|x^\pm - \bar{y}| < C \min\big\{\bar{\sigma}(\varepsilon_0)/2L, |\bar{y} - y_\alpha^{\tilde{\theta}}|\big\}$ such that $\gamma(\theta_\mu)(x^\pm) \to \gamma(\tilde{\theta})(x^\pm)$. Since w-Lip$\big(\gamma(\theta)\big) \le L$ for every θ we obtain that $\gamma(\theta)$ is discontinuous at $\tilde{\theta}$. This gives the contradiction. The proof of the continuity of y_i^θ is entirely similar.
The fact that u^θ solves a quasilinear system guarantees the continuity in t of y_α^θ, y_i^θ, σ_α^θ, and σ_i^θ.
The continuity of σ_α^θ, as a function of θ, follows from the continuity of γ in $\mathbf{L}^1$ and the uniform Lipschitz continuity of the maps $\gamma(\theta)$ for every $\theta \in [0,1]$. Indeed, assume that $\sigma_\alpha^{\theta_\mu} \to \sigma_0 \neq \sigma_\alpha^{\tilde{\theta}}$. There exists $\varepsilon_1 > 0$ such that the sets

$$
\begin{aligned}
A_1 &\doteq \big\{(\exp \sigma r_h)(x) : |x - \gamma(\tilde{\theta})(y_\alpha^{\tilde{\theta}}-)| < \varepsilon_1, |\sigma - \sigma_0| < \varepsilon_1\big\}, \\
A_2 &\doteq \big\{(\exp \sigma r_h)(x) : |x - \gamma(\tilde{\theta})(y_\alpha^{\tilde{\theta}}-)| < \varepsilon_1, |\sigma - \sigma_\alpha^{\tilde{\theta}}| < \varepsilon_1\big\},
\end{aligned}
$$

are disjoint. It is clear that $\gamma(\tilde{\theta})(y_\alpha^{\tilde{\theta}}+) \in A_2$ and there exists $\varepsilon_2 > 0$ such that $B\big(\gamma(\tilde{\theta})(y_\alpha^{\tilde{\theta}}+), \varepsilon_2\big) \subset A_2$ (here $B(y, \rho)$ denotes the ball centered in y with radius ρ). Since $\gamma(\theta)$ is continuous in $\mathbf{L}^1$, there exist $x^- < y_\alpha^{\tilde{\theta}} < x^+$ such that $|x^- - y_\alpha^{\tilde{\theta}}| < \varepsilon_1/6L$, $|x^+ - y_\alpha^{\tilde{\theta}}| < \varepsilon_2/6L$, $\gamma(\theta_\mu)(x^\pm) \to \gamma(\tilde{\theta})(x^\pm)$. Moreover, we can assume that for θ sufficiently close to $\tilde{\theta}$, $|y_\alpha^\theta - y_\alpha^{\tilde{\theta}}| < \min\{\varepsilon_1/6L, \varepsilon_2/6L\}$, and $\gamma(\theta)$ has no shock in $[x^-, x^+] \setminus \{y_\alpha^\theta\}$. From w-Lip$\big(\gamma(\theta)\big) < L$ one has

$$
\big|\gamma(\theta)(x^-) - \gamma(\theta)(y_\alpha^\theta -)\big| < \frac{\varepsilon_1}{3}, \qquad \big|\gamma(\theta)(x^+) - \gamma(\theta)(y_\alpha^\theta +)\big| < \frac{\varepsilon_2}{3}.
$$

Now

$$
\begin{aligned}
\big|\gamma(\theta_\mu)(y_\alpha^{\theta_\mu}-) - \gamma(\tilde{\theta})(y_\alpha^{\tilde{\theta}}-)\big| \le & \big|\gamma(\theta_\mu)(y_\alpha^{\theta_\mu}-) - \gamma(\theta_\mu)(x^-)\big| \\
& + |\gamma(\theta_\mu)(x^-) - \gamma(\tilde{\theta})(x^-)| \\
& + \big|\gamma(\tilde{\theta})(x^-) - \gamma(\tilde{\theta})(y_\alpha^{\tilde{\theta}}-)\big| \\
< & \varepsilon_1,
\end{aligned}
\tag{4.66}
$$

for μ sufficiently large. In the same way it follows

$$
\big|\gamma(\theta_\mu)(y_\alpha^{\theta_\mu}+) - \gamma(\tilde{\theta})(y_\alpha^{\tilde{\theta}}+)\big| < \varepsilon_2. \tag{4.67}
$$

For μ large enough we have $|\sigma_\alpha^{\theta_\mu} - \sigma_0| < \varepsilon_1$, then from (4.66) it follows that $\gamma(\theta_\mu)(y_\alpha^{\theta_\mu}+) \in A_1$. On the other hand from (4.67) we have that $\gamma(\theta_\mu)(y_\alpha^{\theta_\mu}+) \in A_2$, reaching a contradiction.
Finally, the conclusion for the maps $\theta \mapsto \xi_i^\theta$, ξ_α^θ, is provided by Lemma 4.9. □

PROOF OF LEMMA 4.12. Let $[a_k, b_k]$ be a connected component of $[0,1] \setminus J(\varepsilon_0)$. Recall the Remark before Lemma 4.11. From the definition of the ε-accurate Riemann solver in Section 2, if two small h-shocks interact then they simply merge together, without producing outgoing waves of any other family. This means that, if $y_\alpha^\theta(t) = y_\beta^\theta(t)$ for some $\alpha, \beta \in \mathcal{S}$, then $y_\alpha^\theta(s) = y_\beta^\theta(s)$ for every $s \geq \tau$. For every $\alpha, \beta \in \mathcal{S}$ let us define the function

$$f_{\alpha\beta}(\theta) \doteq \sup\left\{t \in [0,T];\ y_\alpha^\theta(t) \neq y_\beta^\theta(t)\right\} \wedge \left(\bar{t} + \tau(L)\right).$$

The maps $f_{\alpha\beta}\colon [a_k, b_k] \to \mathbb{R}$ are Lipschitz continuous. Indeed, fix $\tilde{\theta} \in [a_k, b_k]$ and $\tilde{t} \doteq f_{\alpha\beta}(\tilde{\theta})$. Assume for simplicity that only $y_\alpha^{\tilde{\theta}}$ and $y_\beta^{\tilde{\theta}}$ interact at time $\tilde{t}$. Given $\eta > 0$, there exists $\rho > 0$ such that, at time $\bar{t} = \tilde{t} - \eta$, the shocks y_α^θ and y_β^θ are not coinciding for $\theta \in I_\rho \doteq [\tilde{\theta} - \rho, \tilde{\theta} + \rho]$. Hence the shifts ξ_α^θ, ξ_β^θ, are well defined and continuous for $\theta \in I_\rho$. In particular the maps $\theta \mapsto y_\alpha^\theta$, $\theta \mapsto y_\beta^\theta$, are Lipschitz continuous on I_ρ. Moreover, for every $\theta \in I_\rho$ and $t < f_{\alpha\beta}(\theta)$:

$$\dot{y}_\alpha^\theta(t) - \dot{y}_\beta^\theta(t) \geq \dot{y}_\alpha^{\tilde{\theta}}(t) - \dot{y}_\beta^{\tilde{\theta}}(t) - C|\theta - \tilde{\theta}|.$$

Then, choosing ρ small, for θ sufficiently close to $\tilde{\theta}$ the two shocks will merge within a time

$$t(\theta) \leq \tilde{t} + C|\theta - \tilde{\theta}|,$$

for some $C > 0$. Similarly we obtain $\tilde{t} \leq t(\theta) + C|\theta - \tilde{\theta}|$. The Lipschitz continuity of $f_{\alpha\beta}$ now follows from the compactness of $[a_k, b_k]$.
The set $B(\bar{t})$ is contained in $\cup_{\alpha,\beta} f_{\alpha\beta}^{-1}(\{\bar{t}\})$. By the coarea formula (see [**E-G**]) we have that

$$+\infty > \int_0^1 |f'_{\alpha\beta}(\theta)|\, d\theta = \int_{-\infty}^{+\infty} \mathcal{H}^0\left(f_{\alpha\beta}^{-1}(\{t\})\right) dt,$$

where $\mathcal{H}^0$ is the counting measure. This implies that $\mathcal{H}^0\big(f_{\alpha\beta}^{-1}(\{t\})\big) < +\infty$ for almost every $t \in [0,T]$, hence $\mathcal{H}^0\big(B(t)\big) \leq \sum_{\alpha,\beta} \mathcal{H}^0\big(f_{\alpha\beta}^{-1}(\{t\})\big) < +\infty$ for a.e. t.
It remains to consider the case of interaction of a small shock with a shock layer or with the set (4.6). This can be treated in the same way, using the Lipschitz continuity of the maps $\theta \to y_\alpha^\theta$ and coarea formula. □

PROOF OF LEMMA 4.13. Recall (3.43), (3.45). The map $\theta \mapsto (\gamma(\theta))_x^i$ is continuous in $\mathbf{L}^1$ and a big shock does not interact with any other shock. Hence, from Lemma 4.11 the map $\theta \mapsto R_i^{\gamma(\theta)}(y_i^\theta)$ is continuous except if a small shock enters or exits a shock layer around a big shock. Therefore, it is continuous for every $\theta \notin B(t')$.
Recall (3.32). The map $\theta \mapsto Q(\gamma(\theta))$ suffers a discontinuity only when the configuration changes. Thus it is continuous for every $\theta \in [0,1] \setminus B(t')$. We obtain the conclusion for the map $\theta \mapsto W_i^{\gamma(\theta)}(y_i^\theta)$.
Recall (3.43), (3.44). The map $\theta \mapsto R_h^{\gamma(\theta)}(y_\alpha^\theta)$ is discontinuous only at points $\theta \in B(t')$. Indeed, the first and the third term in the summation (3.44) vary continuously with θ. The second term is discontinuous at θ only if two small shocks

of u^θ interact at time t'. Moreover, the fourth term is discontinuous at θ only if a small shock enters or exits a shock layer around a big shock. Therefore, the first part of the lemma is proved.

Next, for a fixed x, the map $\theta \mapsto W_j^{\gamma(\theta)}(x)$ may suffer a discontinuity in θ only if either $x = y_i^\theta$, $y_i^\theta \pm \delta_2$, for some $i \in \mathcal{S}$ or $x = y_\alpha^\theta$ for some $\alpha \in \mathcal{S}'$. Thus the first conclusion holds. Let $|\mathcal{S}|$ denote the cardinality of the set $\mathcal{S}$. From (3.44) we have

$$\left|R_i^{\gamma(\theta)}(x)\right| \leq V\big(\gamma(\theta)\big) + 2\varepsilon|\mathcal{S}|,$$

while from (3.45) and w-Lip$\big(\gamma(\theta)\big) < L$

$$\left|R_i^{\gamma(\theta)}(y_i)\right| \leq V\big(\gamma(\theta)\big) + 2\varepsilon n L\delta_2,$$

so that

$$\left|W_i^{\gamma(\theta)}(x)\right| \leq 1 + \kappa_1\big(\delta_0 + \varepsilon(2|\mathcal{S}| + 2nL\delta_2)\big) + \kappa_1\kappa_2\delta_0.$$

Therefore, $W_l^{\gamma(\theta)}$ has values in $\mathbf{L}^\infty$. The continuity in $\mathbf{L}^1_{loc}$ follows from the first part and Lebesgue dominated convergence theorem. □

PROOF OF LEMMA 4.14. The vector v^θ satisfies the semilinear system (1.34), with initial data in $\mathbf{L}^\infty$ after each restarting time. Since there is no interaction of big shocks and the interactions of small shocks do not produce new waves, the conclusion follows from [**B-M1**]. □

PROOF OF LEMMA 4.15. The conclusion (i) is obvious and (ii) follows from Lemma 4.11. From the continuity in $\mathbf{L}^1$ of the map $\theta \mapsto (\gamma(\theta))_x^i$, we obtain (iii). Moreover, (v) follows directly from Lemma 4.13.
From Lemma 4.14 and the piecewise Lipschitz continuity of $\gamma(\theta)$, it follows that the map $(\theta, x) \mapsto \gamma(\theta)(x)$ is Lipschitz continuous outside the jumps. Hence (iv) holds. Since all jumps in the functions $\gamma(\theta)$, $\theta \in [\theta_k, \theta_{k+1}]$ are contained in J_k, by the definition of generalized tangent vector we have

$$\begin{aligned}
&\int_{\theta_k}^{\theta_{k+1}} \int_{\mathbb{R}\setminus J_k} \left| \frac{\gamma(\theta_{k+1})(x) - \gamma(\theta_k)(x)}{\theta_{k+1} - \theta_k} - v^\theta(x) \right| dx\, d\theta \\
&\qquad \leq \int_{\theta_k}^{\theta_{k+1}} \int_{\mathbb{R}\setminus J_k} \int_{\theta_k}^{\theta_{k+1}} \left| \frac{v^{\theta'}(x) - v^\theta(x)}{\theta_{k+1} - \theta_k} \right| d\theta'\, dx\, d\theta \\
&\qquad \leq (\theta_{k+1} - \theta_k) \cdot \sup_{\theta, \theta' \in [\theta_k, \theta_{k+1}]} \int_{\mathbb{R}\setminus J_k} \left|v^{\theta'}(x) - v^\theta(x)\right| dx.
\end{aligned}$$

Therefore, (vi) follows from the $\mathbf{L}^1$-continuity of the map $\theta \mapsto v^\theta$.
Finally, from Lemmas 4.11 and 4.13, we obtain (vii). □

PROOF OF LEMMA 4.16. The proof of (4.32) follows directly from Lemma 4.3. Indeed, the estimate (4.61) depends only on the weighted Lipschitz constant L and on the choice of the step δ_4, and hence it is uniform in θ.
From (4.62) and (4.63) we have

$$\tilde{u}_x = r_h(U)[u_x^h]_+ + \sum_{j \neq h} r_j(U) u_x^j + O(\delta_4).$$

This implies

$$\begin{aligned}\tilde{u}_x^j =&\langle l_j(\tilde{u}), \tilde{u}_x\rangle = \\ =&\langle l_j(\tilde{u}),\ r_h(\tilde{u})[u_x^h]_+ + \sum_{j\neq h} r_j(\tilde{u})u_x^j\rangle + \langle l_j(\tilde{u}), (r_h(U) - r_h(\tilde{u}))[u_x^h]_+\rangle \\ &+ \sum_{j\neq h}\Big\langle l_j(\tilde{u}), \big(r_j(U) - r_j(\tilde{u})\big)u_x^j\Big\rangle + O(\delta_4).\end{aligned}$$

For $j \neq h$, the first term of the last expression is u_x^j, hence

$$|u_x^j - \tilde{u}_x^j| \leq CL|\tilde{u} - U| + C\delta_4.$$

Recalling that $\gamma(\theta)$ and $P_{\delta_4}\big(\gamma(\theta)\big)$ coincide outside $[-M_0, M_0]$, (4.33) follows. The estimate (4.34) follows from the analogous one in Lemma 4.3. □

PROOF OF LEMMA 4.19. On the set $\mathcal{A}$ we have

$$\begin{aligned}\Big|P_{\delta_4}&\big(\gamma(\theta_k)\big)(x) - \tilde{\gamma}(\theta)(x)\Big| \\ &\leq C\Big|c_k^\theta\Phi\big(P_{\delta_4}(\gamma(\theta_k))(x)\big) \\ &\quad + (1 - c_k^\theta)\Phi\big(P_{\delta_4}(\gamma(\theta_{k+1}))(x)\big) - \Phi\big(P_{\delta_4}(\gamma(\theta_k))(x)\big)\Big| \\ &\leq C\Big|\Phi\big(P_{\delta_4}(\gamma(\theta_{k+1}))(x)\big) - \Phi\big(P_{\delta_4}(\gamma(\theta_k))(x)\big)\Big| \\ &\leq C\Big\{\big|P_{\delta_4}(\gamma(\theta_{k+1}))(x) - \gamma(\theta_{k+1})(x)\big| \\ &\quad + \big|P_{\delta_4}(\gamma(\theta_k))(x) - \gamma(\theta_k)(x)\big| + \big|\gamma(\theta_{k+1})(x) - \gamma(\theta_k)(x)\big|\Big\} \\ &\leq 3C\varepsilon_0,\end{aligned} \tag{4.68}$$

where $c_k^\theta \doteq (\theta_{k+1} - \theta)/(\theta_{k+1} - \theta_k)$. The last inequality follows from (4.32) and (4.25). From (4.68), using again (4.32) and (4.25) one obtains

$$\begin{aligned}\big|\gamma(\theta)(x) - \tilde{\gamma}(\theta)(x)\big| \leq&\big|\gamma(\theta)(x) - \gamma(\theta_k)(x)\big| + \big|\gamma(\theta_k)(x) - P_{\delta_4}(\gamma(\theta_k))(x)\big| \\ &+ \big|P_{\delta_4}(\gamma(\theta_k))(x) - \tilde{\gamma}(\theta)(x)\big| \\ \leq& C\varepsilon_0,\end{aligned}$$

which proves the lemma. □

PROOF OF LEMMA 4.20. Defining

$$y(\theta, x) \doteq c_k^\theta\Phi\big(P_{\delta_4}(\gamma(\theta_k))(x)\big) + (1 - c_k^\theta)\Phi\big(P_{\delta_4}(\gamma(\theta_{k+1}))(x)\big), \tag{4.69}$$

we have

$$\begin{aligned}\big(\tilde{\gamma}(\theta)\big)_x(x) =& D\Phi^{-1}(y(\theta, x)) \cdot \Big[c_k^\theta D\Phi\big(P_{\delta_4}(\gamma(\theta_k))(x)\big) \cdot \big(P_{\delta_4}(\gamma(\theta_k))\big)_x(x) \\ &+ (1 - c_k^\theta)D\Phi\big(P_{\delta_4}(\gamma(\theta_{k+1}))(x)\big) \cdot \big(P_{\delta_4}(\gamma(\theta_{k+1}))\big)_x(x)\Big].\end{aligned}$$

Hence, from the proof of Lemma 4.19 it follows

$$
\begin{aligned}
&\left|(\tilde{\gamma}(\theta))^j_x(x) - (\gamma(\theta))^j_x(x)\right| \leq \\
&\leq \left|\left(c^\theta_k (P_{\delta_4}(\gamma(\theta_k)))^j_x(x) + (1-c^\theta_k)(P_{\delta_4}(\gamma(\theta_{k+1})))^j_x(x) - (\gamma(\theta))^j_x(x)\right)\right| \\
&\quad + C\left|D\Phi^{-1}(y(\theta,x)) - D\Phi^{-1}\left[\Phi(P_{\delta_4}(\gamma(\theta_k))(x))\right]\right| \\
&\quad + C\left|D\Phi^{-1}(y(\theta,x)) - D\Phi^{-1}\left[\Phi(P_{\delta_4}(\gamma(\theta_{k+1}))(x))\right]\right| \\
&\quad + C\left(\left|\tilde{\gamma}(\theta)(x) - P_{\delta_4}(\gamma(\theta_k))(x)\right| + \left|\tilde{\gamma}(\theta)(x) - P_{\delta_4}(\gamma(\theta_{k+1}))(x)\right|\right) \\
&\quad + C\left|\tilde{\gamma}(\theta)(x) - \gamma(\theta)(x)\right| \\
&\leq c^\theta_k \left|(P_{\delta_4}(\gamma(\theta_k)))^j_x(x) - (\gamma(\theta_k))^j_x(x)\right| \\
&\quad + (1-c^\theta_k)\left|(P_{\delta_4}(\gamma(\theta_{k+1})))^j_x(x) - (\gamma(\theta_{k+1}))^j_x(x)\right| \\
&\quad + c^\theta_k \left|(\gamma(\theta_k))^j_x(x) - (\gamma(\theta))^j_x(x)\right| + (1-c^\theta_k)\left|(\gamma(\theta_{k+1}))^j_x(x) - (\gamma(\theta))^j_x(x)\right| \\
&\quad + C\left|P_{\delta_4}(\gamma(\theta_k))(x) - P_{\delta_4}(\gamma(\theta_{k+1}))(x)\right| + C\varepsilon_0 \\
&\doteq I_1(x) + I_2(x) + I_3(x) + I_4(x) + I_5(x) + C\varepsilon_0.
\end{aligned} \tag{4.70}
$$

In the following we use $\|\cdot\|_{\mathbf{L}^1}$ to indicate the $\mathbf{L}^1$ norm restricted to $\mathcal{A}$. From (4.33) of Lemma 4.16 we have

$$\|I_1\|_{\mathbf{L}^1} + \|I_2\|_{\mathbf{L}^1} \leq \varepsilon_0. \tag{4.71}$$

Using (4.24) and Lemma 4.15 one obtains

$$\|I_3\|_{\mathbf{L}^1} + \|I_4\|_{\mathbf{L}^1} \leq \varepsilon_0. \tag{4.72}$$

while, by Lemmas 4.16 and 4.15 one gets

$$\|I_5\|_{\mathbf{L}^1} \leq C\|\gamma(\theta_k) - \gamma(\theta_{k+1})\|_{\mathbf{L}^1} + 2\varepsilon_0 \leq C\varepsilon_0. \tag{4.73}$$

□

Proof of Lemma 4.21. We fix $k \notin K$ and $\theta \in [\theta_k, \theta_{k+1}]$. Let us denote $u \doteq \gamma(\theta)$, $\tilde{u} \doteq \tilde{\gamma}(\theta)$, and let $\mathcal{S}'$, $\tilde{\mathcal{S}}'$, denote respectively the set of small h-shocks of u and $\tilde{u}$. Moreover, let y_α, $\alpha \in \mathcal{S}'$, (resp. $\tilde{y}_\alpha$, $\alpha \in \tilde{\mathcal{S}}'$) be the locations of these shocks, and σ_α (resp. $\tilde{\sigma}_\alpha$) their strengths. By construction, we clearly have $\mathcal{S}' \subset \tilde{\mathcal{S}}'$. For notational convenience, we let $\mathcal{S}' = \tilde{\mathcal{S}}'$ defining $\sigma_\alpha = 0$, $y_\alpha = \tilde{y}_\alpha$ if $\alpha \in \tilde{\mathcal{S}}' \setminus \mathcal{S}'$. Notice that, for every α, we have

$$|\sigma_\alpha - \tilde{\sigma}_\alpha| < C\delta_4. \tag{4.74}$$

Recall (3.43), (3.44) and (3.32). We first consider the terms R^u_j. The third addendum of (3.44) is clearly the same for u and $\tilde{u}$. While, if $y^u_i \neq y^{\tilde{u}}_i$, then the last addendum is different for some x. However, this can happen only on a set whose measure is bounded by $2\sup_{\theta\in[\theta_k,\theta_{k+1}]} |y^\theta_i - \tilde{y}^\theta_i| \leq C\varepsilon_0\delta_3$ (here $\tilde{y}^\theta_i$ denote the locations of big shocks of $\tilde{\gamma}(\theta)$ and the last inequality is guaranteed by (4.23) of Lemma 4.15).

Let us now consider the first two terms of (3.44). If $j = h$ then, from Lemma 4.20, we obtain

$$|R^u_h(x) - R^{\tilde{u}}_h(x)| \leq C(\varepsilon_0 + \delta_4).$$

Indeed, in this case, the sum in the second term is restricted to the set $\mathcal{S}$. Assume now $j < h$, being similar the other case. Let us define the set $Z(x) \doteq \left\{y \in \cup_{k,i}[y^-_{\bar{k},i}, y^+_{\bar{k},i}];\ y < x\right\}$. We have

$$\int_{Z(x)} \big||\tilde{u}^j_x| - |u^j_x|\big|\, dy < C(\delta_3 + \varepsilon_0).$$

Recall the construction in the proof of Lemma 4.3 and consider an interval $[x^l_r, x^l_{r+1}]$ to the left of x. Let $\beta = \beta(l,r)$ be such that $\tilde{y}_\beta = x^l_{r+1}$. From Lemma 4.3 we deduce that

$$\left|\int_{x^l_r}^{x^l_{r+1}} \left([u^h_x]_- + \tilde{\sigma}_\beta - \sigma_\beta\right) dx\right| \leq C\delta_4. \tag{4.75}$$

We remark that there is at most one couple (l,r) such that $x \in]x^l_r, x^l_{r+1}[$, and the corresponding integral term is estimated by $L\delta_4$. For the h-waves, summing over all (l,r) and using the estimate of Lemma 4.20, we obtain

$$\left|\int_{-\infty}^{x} \left|\tilde{u}^j_x(y)\right| dy + \sum_{\substack{\alpha \in \tilde{\mathcal{S}}' \\ \tilde{y}_\alpha < x}} |\tilde{\sigma}_\alpha| - \int_{-\infty}^{x} \left|u^j_x(y)\right| dy - \sum_{\substack{\alpha \in \mathcal{S}' \\ y_\alpha < x}} |\sigma_\alpha|\right| \leq C\varepsilon_0. \tag{4.76}$$

Using again Lemma 4.20, we can treat the first term of (3.44) for the other family of waves, obtaining

$$\left|R_j^{\gamma(\theta_k)}(x) - R_j^{\tilde{\gamma}(\theta_k)}(x)\right| < C(\delta_3 + \delta_4 + \varepsilon_0).$$

We now work toward an estimate on Q. We use again the notations of Lemma 4.3. Let us consider two intervals $[x^l_r, x^l_{r+1}]$, $[x^m_s, x^m_{s+1}]$ satisfying $x^l_{r+1} \leq x^m_s$. Let $\beta = \beta(l,r)$ be such that $\tilde{y}_\beta = x^l_{r+1}$. Let us define

$$E \doteq \int_{x^l_r}^{x^l_{r+1}} \left|u^h_x(y)\right| dy \int_{x^m_s}^{x^m_{s+1}} \left|u^j_x(y)\right| dy + |\sigma_\beta| \int_{x^m_s}^{x^m_{s+1}} \left|u^j_x(y)\right| dy,$$

and let $\tilde{E}$ be the corresponding quantity for $\tilde{u}$. By Lemma 4.20, and the above arguments, we have the estimates

$$\begin{aligned}
I_1 \doteq& \int_{x^l_r}^{x^l_{r+1}} \left|u^h_x(y)\right| dy \left|\int_{x^m_s}^{x^m_{s+1}} \left(\left|u^j_x(y)\right| - \left|\tilde{u}^j_x(y)\right|\right) dy\right| \\
<& C(\varepsilon_0 + \delta_4) \int_{x^l_r}^{x^l_{r+1}} \left|u^h_x(y)\right| dy, \\
I_2 \doteq& \left|\int_{x^l_r}^{x^l_{r+1}} \Big|\left|u^h_x(y)\right| - \left|\tilde{u}^h_x(y)\right|\Big| dy \int_{x^m_s}^{x^m_{s+1}} \left|u^j_x(y)\right| dy + |\sigma_\beta| \int_{x^m_s}^{x^m_{s+1}} \left|u^j_x(y)\right| dy\right| \\
\leq& C(\varepsilon_0 + \delta_4) \int_{x^m_s}^{x^m_{s+1}} \left|u^j_x(y)\right| dy, \\
I_3 \doteq& |\sigma_\beta| \int_{x^m_s}^{x^m_{s+1}} \Big|\left|u^j_x(y)\right| - \left|\tilde{u}^j_x(y)\right|\Big| dy \\
<& C\delta_0(\varepsilon_0 + \delta_4)|\sigma_\beta|.
\end{aligned}$$

We thus obtain

$$|E-\tilde{E}| \leq I_1 + I_2 + I_3 \leq C(\varepsilon_0+\delta_4)\text{T.V.}\{u, [x_r^l, x_{r+1}^l] \cup [x_s^m, x_{s+1}^m]\}.$$

We then sum over all pairs of intervals.
The case of approaching j-waves and k-waves, with $j,k \neq h$ is easily treated using Lemma 4.20. Summing over all the intervals we can estimate the first two terms in the expression (3.32) for Q.
If $\alpha,\beta \in \mathcal{S}\cup\mathcal{S}'$, by (4.29) of Lemma 4.15, one obtains

$$\Big||\tilde{\sigma}_\alpha^\theta \tilde{\sigma}_\beta^\theta| - |\sigma_\alpha^\theta \sigma_\beta^\theta|\Big| \leq \delta_0\left(|\tilde{\sigma}_\alpha^\theta - \sigma_\alpha^\theta| + |\tilde{\sigma}_\beta^\theta - \sigma_\beta^\theta|\right) \leq 2\delta_0\varepsilon_0.$$

Again the other terms are estimated using Lemma 4.20 and (4.74). We thus obtain

$$\Big|Q\big(\gamma(\theta)\big) - Q\big(\tilde{\gamma}(\theta)\big)\Big| < C(\varepsilon_0+\delta_4).$$

Recalling (3.44), from the estimates on R_j^u and Q the conclusion follows. □

PROOF OF LEMMA 4.22. Fix $k \notin K$, consider the set $\mathcal{A}_k \doteq \big([-M_0,M_0]\times[\theta_k,\theta_{k+1}]\big)\cap(\mathcal{A}\setminus J_k)$ (see (iv) of Lemma 4.15 for the definition of J_k) and let $\tilde{v}$ be the tangent vector associated to $\tilde{\gamma}$ on $\mathcal{A}_k$. We have

$$\begin{aligned}\tilde{v}^\theta =& D\Phi^{-1}\Big(c_k^\theta \Phi\big(P_{\delta_4}(\gamma(\theta_k))\big) + (1-c_k^\theta)\Phi\big(P_{\delta_4}(\gamma(\theta_{k+1}))\big)\Big)\cdot \\ &\cdot\left(\frac{\Phi\big(P_{\delta_4}(\gamma(\theta_k))\big) - \Phi\big(P_{\delta_4}(\gamma(\theta_{k+1}))\big)}{\theta_{k+1}-\theta_k}\right)\end{aligned}$$

on the set $\mathcal{A}_k$. For every i, we have to estimate the quantity

$$\begin{aligned}&\left|\iint_{\mathcal{A}_k}\Big\{|\langle l_j(\gamma(\theta)), v^\theta\rangle| W_j^{\gamma(\theta)} - |\langle l_j(\tilde{\gamma}(\theta)), \tilde{v}^\theta\rangle| W_j^{\tilde{\gamma}(\theta)}\Big\}\,dx\,d\theta\right| \\ &\quad\leq\left|\iint_{\mathcal{A}_k}|\langle l_j(\gamma(\theta)), v^\theta\rangle|\big(W_j^{\gamma(\theta)} - W_j^{\tilde{\gamma}(\theta)}\big)\,dx\,d\theta\right| \\ &\qquad+\left|\iint_{\mathcal{A}_k}\big(|\langle l_j(\gamma(\theta)), v^\theta\rangle| - |\langle l_j(\tilde{\gamma}(\theta)), v^\theta\rangle|\big)W_j^{\tilde{\gamma}(\theta)}\,dx\,d\theta\right| \\ &\qquad+\left|\iint_{\mathcal{A}_k}\big(|\langle l_j(\tilde{\gamma}(\theta)), v^\theta\rangle| - |\langle l_j(\tilde{\gamma}(\theta)), \tilde{v}^\theta\rangle|\big)W_j^{\tilde{\gamma}(\theta)}\,dx\,d\theta\right| \\ &\quad\doteq I_1 + I_2 + I_3.\end{aligned}$$

Let us first prove that

$$\iint_{\mathcal{A}_k}\left|\frac{\gamma(\theta_{k+1})(x) - \gamma(\theta_k)(x)}{\theta_{k+1}-\theta_k} - \tilde{v}^\theta(x)\right|\,dx\,d\theta < C(\varepsilon_0+\delta_4)(\theta_{k+1}-\theta_k). \tag{4.77}$$

From (4.69), we have

$$\begin{aligned}
&\int_{\mathcal{A}_k} \left| \frac{\gamma(\theta_{k+1})(x) - \gamma(\theta_k)(x)}{\theta_{k+1} - \theta_k} - \tilde{v}^\theta(x) \right| dx\, d\theta \\
&= \int_{\mathcal{A}_k} \Bigg| \frac{\gamma(\theta_{k+1})(x) - \gamma(\theta_k)(x)}{\theta_{k+1} - \theta_k} \\
&\quad - D\Phi^{-1}(y(\theta,x)) \cdot \left(\frac{\Phi\big(P_{\delta_4}(\gamma(\theta_k))(x)\big) - \Phi\big(P_{\delta_4}(\gamma(\theta_{k+1}))(x)\big)}{\theta_{k+1} - \theta_k} \right) \Bigg| \, dx\, d\theta \\
&\leq \int_{\mathcal{A}_k} \frac{\left|\gamma(\theta_{k+1})(x) - P_{\delta_4}(\gamma(\theta_{k+1}))(x)\right| + \left|\gamma(\theta_k)(x) - P_{\delta_4}(\gamma(\theta_k))(x)\right|}{|\theta_{k+1} - \theta_k|} \, dx\, d\theta \\
&\quad + C\|v^\theta\|_{\mathbf{L}^\infty} \int_{\mathcal{A}_k} \left| D\Phi^{-1}(y(\theta,x)) - D\Phi^{-1}\left(\Phi\left(P_{\delta_4}(\gamma(\theta_k))(x)\right)\right) \right| dx\, d\theta \\
&\quad + C\|v^\theta\|_{\mathbf{L}^\infty} \int_{\mathcal{A}_k} \left| D\Phi^{-1}(y(\theta,x)) - D\Phi^{-1}\left(\Phi\left(P_{\delta_4}(\gamma(\theta_{k+1}))(x)\right)\right) \right| dx\, d\theta \\
&\quad + C\varepsilon_0 \|v^\theta\|_{\mathbf{L}^\infty}.
\end{aligned}$$

The first integral is estimated by $C\varepsilon_0(\theta_{k+1} - \theta_k)$, using (4.32). The integrands of the last two integrals are estimated, using (4.32) and (4.25) of Lemma 4.15, by

$$\begin{aligned}
C\Big|P_{\delta_4}\big(\gamma(\theta_k)\big)(x) - P_{\delta_4}\big(\gamma(\theta_{k+1})\big)(x)\Big| &\leq C\varepsilon_0 + C\left|\gamma(\theta_k)(x) - \gamma(\theta_{k+1})(x)\right| \\
&\leq C(\varepsilon_0 + \delta_4).
\end{aligned}$$

Hence the last two integrals are estimated by $C\, M_0\, (\varepsilon_0 + \delta_4)(\theta_{k+1} - \theta_k)$.
From Lemma 4.21 the following inequality holds:

$$\left| \int_{\mathcal{A}_k} \left[W_j^{\gamma(\theta)}(x) - W_j^{\tilde{\gamma}(\theta)}(x) \right] dx\, d\theta \right| < C(\varepsilon_0 + \delta_3 + \delta_4)(\theta_{k+1} - \theta_k). \tag{4.78}$$

Now, from Lemma 4.14 and (4.78) one has

$$\begin{aligned}
I_1 \leq & C\|v^\theta\|_{\mathbf{L}^\infty} \left| \int_{\mathcal{A}_k} W_j^{\gamma(\theta)}(x) - W_j^{\tilde{\gamma}(\theta)}(x)\, dx\, d\theta \right| \\
< & C\|v^\theta\|_{\mathbf{L}^\infty} (\varepsilon_0 + \delta_3 + \delta_4)(\theta_{k+1} - \theta_k).
\end{aligned}$$

From Lemma 4.19

$$\begin{aligned}
I_2 \leq & C \int_{\mathcal{A}_k} \Big(|\langle l_j(\gamma(\theta)), v^\theta\rangle - \langle l_j(\tilde{\gamma}(\theta)), v^\theta\rangle| \Big)\, dx\, d\theta \\
\leq & C\|v^\theta\|_{\mathbf{L}^\infty} \int_{\mathcal{A}_k} \left|\gamma(\theta)(x) - \tilde{\gamma}(\theta)(x)\right| dx\, d\theta < \\
< & C\|v^\theta\|_{\mathbf{L}^\infty} (\varepsilon_0 + \delta_4)(\theta_{k+1} - \theta_k).
\end{aligned}$$

Finally, from (4.28) of Lemma 4.15 and (4.77) we get

$$\begin{aligned} I_3 \leq & C \int_{\mathcal{A}_k} |l_j(\tilde{\gamma}(\theta))| \, |v^\theta - \tilde{v}^\theta| \, dx \, d\theta \\ \leq & C \int_{\mathcal{A}_k} \left| v^\theta(x) - \frac{\gamma(\theta_{k+1})(x) - \gamma(\theta_{k+1})}{\theta_{k+1} - \theta_k} \right| dx \, d\theta \\ & + C \int_{\mathcal{A}_k} \left| \frac{\gamma(\theta_{k+1})(x) - \gamma(\theta_{k+1})}{\theta_{k+1} - \theta_k} - \tilde{v}^\theta(x) \right| dx \, d\theta \\ < & C(\varepsilon_0 + \delta_4)(\theta_{k+1} - \theta_k). \end{aligned}$$

Now, since $meas(J_k) \leq \varepsilon_0 \Delta t(\theta_{k+1} - \theta_k)$, summing over k we obtain the conclusion. □

Proof of Lemma 4.23. From the proofs of Lemma 4.3 and Lemma 4.16 we obtain, for every $i = 1, \ldots, n$ and every k,

$$\left| (\tilde{\gamma}(\theta_k))_x^j - (\gamma(\theta_k))_x^j \right| \leq C(\varepsilon_0 + \delta_4), \qquad \left| W_j^{\tilde{\gamma}(\theta_k)} - W_j^{\gamma(\theta_k)} \right| \leq C(\varepsilon_0 + \delta_4). \tag{4.79}$$

These estimates imply

$$\begin{aligned} & \left| |(\tilde{\gamma}(\theta_k))_x^j(x)| e^{\beta W_j^{\tilde{\gamma}(\theta_k)}(x)} - |(\gamma(\theta_k))_x^j(x)| e^{\beta W_j^{\gamma(\theta_k)}(x)} \right| \\ & \qquad \leq C \left| (\tilde{\gamma}(\theta_k))_x^j(x) - (\gamma(\theta_k))_x^j(x) \right| + C e^{\beta \left| W_j^{\tilde{\gamma}(\theta_k)}(x) - W_j^{\gamma(\theta_k)}(x) \right|} \\ & \qquad \leq C(\varepsilon_0 + \delta_4). \end{aligned}$$

From (4.36), on the set $\mathcal{A}$, $(\tilde{\gamma}(\theta))_x(x)$ is expressed as in Lemma 4.20. We can estimate $(\tilde{\gamma}(\theta))_x$ as in (4.70) obtaining

$$\begin{aligned} \left| (\tilde{\gamma}(\theta))_x(x) \right| \leq & \left| c_k^\theta \left[P_{\delta_4}(\gamma(\theta_k)) \right]_x(x) + (1 - c_k^\theta) \left[P_{\delta_4}(\gamma(\theta_{k+1})) \right]_x(x) \right| \\ & + C \left| D\Phi^{-1}(y(\theta, x)) - D\Phi^{-1}(\Phi(P_{\delta_4}(\gamma(\theta_k))(x))) \right| \\ & + C \left| D\Phi^{-1}(y(\theta, x)) - D\Phi^{-1}\Phi(P_{\delta_4}(\gamma(\theta_{k+1}))(x)) \right| \\ \leq & c_k^\theta \left| (\gamma(\theta_k))_x(x) \right| + (1 - c_k^\theta) \left| (\gamma(\theta_{k+1}))_x(x) \right| + C(\varepsilon_0 + \delta_4). \end{aligned} \tag{4.80}$$

Let us now first consider the points x for which it holds the conclusion of Lemma 4.21. Then, by Lemma 4.21, Lemma 4.15(v) and (4.79) we have

$$\begin{aligned} \left| W_j^{\tilde{\gamma}(\theta)}(x) - W_j^{\tilde{\gamma}(\theta_{k+1})}(x) \right| \leq & \left| W_j^{\tilde{\gamma}(\theta)}(x) - W_j^{\gamma(\theta)}(x) \right| \\ & + \left| W_j^{\gamma(\theta)}(x) - W_j^{\gamma(\theta_k)}(x) \right| \\ & + \left| W_j^{\gamma(\theta_k)}(x) - W_j^{\tilde{\gamma}(\theta_{k+1})}(x) \right| \\ \leq & C(\varepsilon_0 + \delta_4). \end{aligned} \tag{4.81}$$

Hence by Lemma 4.15, (4.80) and (4.81) we obtain

$$\begin{aligned}
&\Big| |(\tilde\gamma(\theta))^j_x(x)| e^{\beta W_j^{\tilde\gamma(\theta)}(x)} - c^\theta_k |(\tilde\gamma(\theta_k))^j_x(x)| e^{\beta W_j^{\tilde\gamma(\theta_k)}(x)} \\
&\qquad - (1-c^\theta_k) |(\tilde\gamma(\theta_{k+1}))^j_x(x)| e^{\beta W_j^{\tilde\gamma(\theta_{k+1})}(x)} \Big| \\
&\leq C \Big\{ c^\theta_k |(\tilde\gamma(\theta_k))^j_x(x)| e^{\beta\left(W_j^{\tilde\gamma(\theta)}(x) - W_j^{\tilde\gamma(\theta_k)}(x)\right)} \\
&\qquad + (1-c^\theta_k) |(\tilde\gamma(\theta_{k+1}))^j_x(x)| e^{\beta\left(W_j^{\tilde\gamma(\theta)}(x) - W_j^{\tilde\gamma(\theta_{k+1})}(x)\right)} \Big\} + C(\varepsilon_0+\delta_4) \\
&\leq C(\varepsilon_0+\delta_4).
\end{aligned}$$

Therefore we obtain the conclusion.
Consider now a point x for which the conclusion of Lemma 4.21 does not holds. From the proof of Lemma 4.21, these are precisely the points x that lie inside a shock layer of a big shock for $\gamma(\theta)$ but not for $\tilde\gamma(\theta)$. We can assume, for example, that $y^\theta_i - \delta_2 < x < \tilde y^\theta_i - \delta_2$ for some $i \in \mathcal{S}$ (here $\tilde y^\theta_i$, $i \in \mathcal{S}$, denote the positions of big shocks of $\tilde\gamma(\theta)$). In this case

$$W_j^{\tilde\gamma(\theta)}(x) \geq W_j^{\gamma(\theta)}(x) + \varepsilon - C(\varepsilon_0+\delta_4),$$

and the conclusion may fail. However, it is sufficient to modify the values of $\tilde\gamma(\theta_k)$ in such a way that near the points $y^\theta_i \pm \delta_2$, $i \in \mathcal{S}$, the quantities $|(\gamma(x))^i_x|$ are smaller then $L/\exp[\beta\varepsilon]$. This can be done shifting some waves as in Lemma 4.3 or as in Lemma 4.4. Since, by Lemma 4.15(ii), the waves that should be shifted are estimated by $L\varepsilon_0\delta_3$, all the conclusions of the previous Lemmas still hold. □

PROOF OF LEMMA 4.24. Fix k such $]\theta_k, \theta_{k+1}[\subset \mathcal{K}$. We assume that there exists a unique $\theta' \in B(t')$ such that $\theta' \in]\theta_k, \theta_{k+1}[$. We distinguish two cases:

a) $\theta' \in \Theta$;
b) $\theta' \in B(t') \setminus \Theta$.

First consider the case a). We possibly have some small shocks y^θ_α whose strengths go to zero as θ tends to θ'. These come precisely from the interpolation (4.36) implemented at some restarting before t'. Indeed, it may happen that $\gamma(\theta_k)$ presents a shock at a point where $\gamma(\theta_{k+1})$ has no shock, so that the strength of this shock tends to zero as θ tends to θ_{k+1}. Let $\tilde{\mathcal{S}} \subset \mathcal{S}'$ be the set of such shocks. If we neglect these small shocks, the estimates of Lemma 4.15 are still valid separately on the two intervals $]\theta_k, \theta'[$, $]\theta', \theta_{k+1}[$. On these two intervals, we can perform the same construction used on the intervals contained in $[0,1] \setminus \mathcal{K}$, neglecting the shocks y^θ_α, $\alpha \in \tilde{\mathcal{S}}$. Possibly shrinking ε_0, we can let $|\sigma^\theta_\alpha|$, $\alpha \in \tilde{\mathcal{S}}$, be arbitrarily small obtaining the conclusion.

Now consider the case b). The case of interaction of a small shock with a shock layer or with the set (4.6), can be easily treated following the construction of $\tilde\gamma$ on $[0,1] \setminus \mathcal{K}$. Without loss of generality we can assume that there is a simple shock bifurcation at θ'. There exist $\alpha, \beta \in \mathcal{S}'$ such that $\tilde y^{\theta_k}_\alpha < \tilde y^{\theta_k}_\beta$, $\tilde y^{\theta_{k+1}}_\alpha = \tilde y^{\theta_{k+1}}_\beta$. We first modify $\tilde\gamma(\theta_k)$, interpolating linearly the values $\tilde\gamma(\theta_k; \tilde y^{\theta_k}_\alpha +)$, $\tilde\gamma(\theta_k; \tilde y^{\theta_k}_\beta -)$ on the set $[\tilde{y}_\alpha^{\theta_k}, \tilde{y}_\beta^{\theta_k}]$. Let us define the curve γ_1 as follows. Let $d \doteq \tilde{y}_\beta^{\theta_k} - \tilde{y}_\alpha^{\theta_k}$. For $\theta \in [0,1[$ we set

$$\gamma_1(\theta;x) \doteq \begin{cases} \tilde{\gamma}(\theta_k;x) & \text{if } x \in]-\infty, \tilde{y}_\alpha^{\theta_k}[\cup]\tilde{y}_\alpha^{\theta_k} + \theta d, +\infty[\\ (1-\frac{c}{\theta d})\tilde{\gamma}(\theta_k)(\tilde{y}_\alpha^{\theta_k}) + & \\ +\frac{c}{\theta d}(\exp(\tilde{\sigma}_\alpha^{\theta_k} r_h)(\tilde{\gamma}(\theta_k;\tilde{y}_\alpha^{\theta_k}+\theta d)) & \text{if } x = \tilde{y}_\alpha^{\theta_k} + c, \quad c \in [0,\theta d], \end{cases}$$

while for $\theta = 1$ we set

$$\gamma_1(1;x) \doteq \begin{cases} \tilde{\gamma}(\theta_k;x), & \text{if } x \in]-\infty, \tilde{y}_\alpha^{\theta_k}[\cup]\tilde{y}_\beta^{\theta_k}, +\infty[, \\ (1-\frac{c}{d})\tilde{\gamma}(\theta_k)(\tilde{y}_\alpha^{\theta_k}) + & \\ +\frac{c}{d}\exp((\tilde{\sigma}_\alpha^{\theta_k} + \tilde{\sigma}_\beta^{\theta_k})r_h)(\tilde{\gamma}(\theta_k;\tilde{y}_\beta^{\theta_k}+)) & \text{if } x = \tilde{y}_\alpha^{\theta_k} + c, \, c \in [0,d]. \end{cases}$$

Since we shift a shock of strength $\tilde{\sigma}_\alpha^{\theta_k}$ for a length d, and the waves in $[\tilde{y}_\alpha^{\theta_k}, \tilde{y}_\beta^{\theta_k}]$ are estimated by $(L + C\varepsilon_0)d$, we obtain

$$\|\gamma_1\|_\star \leq Cd(\tilde{\sigma}_\alpha^{\theta_k} + L + \varepsilon_0).$$

Now we have that $d = O(\varepsilon_0)$, hence $\|\gamma_1\|_\star = O(\varepsilon_0)$. Moreover, reasoning as in Lemma 4.3, it follows that $\gamma_1(\theta) \in \mathcal{D}^*_{h,L+C\varepsilon_0}$, and it is easy to check that γ_1 is a regular path.
Since $\gamma_1(1)$ and $\tilde{\gamma}(\theta_{k+1})$ have the same number of distinct shocks, then they can be joined, using the construction implemented on $([0,1] \setminus \mathcal{K}) \times [-M_0, M_0]$, by a regular path whose weighted length is still $O(\varepsilon_0)$. The other conclusions are easily verified. □

PROOF OF CLAIM 4.25. Notice that the set

$$\mathcal{Z} \doteq \Big(\bigcup_{k \notin K} [\theta_k, \theta_{k+1}] \times [-M_0, M_0]\Big) \setminus \mathcal{A}$$

can be chosen of arbitrarily small measure, and the curves γ_k of Lemma 4.24 of arbitrarily small length, letting δ_3 and ε_0 small. Moreover, for the set $\mathcal{A}$ we have the estimate given in Lemma 4.22. Now, from Lemma 4.15(vii), for every $\alpha \in \mathcal{S}$

$$\left| \int_{\theta_k}^{\theta_{k+1}} \xi_\alpha^\theta \sigma_\alpha^\theta W_h^{\gamma(\theta)}(y_\alpha^\theta)d\theta - \int_{\theta_k}^{\theta_{k+1}} \tilde{\xi}_\alpha^\theta \tilde{\sigma}_\alpha^\theta W_h^{\tilde{\gamma}(\theta)}(\tilde{y}_\alpha^\theta)d\theta \right| \leq$$

$$\leq \varepsilon_0(\theta_{k+1} - \theta_k) + C\int_{\theta_k}^{\theta_{k+1}} \left| W_h^{\gamma(\theta_k)}(y_\alpha^{\theta_k}) - W_h^{\tilde{\gamma}(\theta)}(\tilde{y}_\alpha^\theta) \right| d\theta.$$

Reasoning as in the proof of Lemma 4.21 and using Lemma 4.15(v), we obtain

$$\left| W_h^{\gamma(\theta_k)}(y_\alpha^{\theta_k}) - W_h^{\tilde{\gamma}(\theta)}(\tilde{y}_\alpha^\theta) \right| \leq C(\varepsilon_0 + \delta_4).$$

The argument can be repeated for every $i \in \mathcal{S}$. Notice that the new small shocks of $\tilde{\gamma}$ are not shifted, hence they give no contribution to the norm of the tangent vector to $\tilde{\gamma}$. Thus, from (3.42) we obtain (4.18). The estimate (4.19) is easily verified using Lemmas 4.15 and 4.24.
The curve $\tilde{\gamma}$ is piecewise regular. Indeed, on the sets $[\theta_k, \theta_{k+1}]$, $k \notin K$, it is defined via suitable interpolations. Hence, $\tilde{\gamma}$ is regular on $]\theta_k, \theta_{k+1}[$, $k \notin K$. It may happen that $(\tilde{v}, \tilde{\xi})$ is not defined at some θ_k, because the strengths of some small shocks (generated by P_{δ_4}) may tend to zero letting the set $\mathcal{S}'$ be discontinuous at θ_k. Since there is a finite number of θ_k's, the conclusion follows from Lemma 4.24. □

PROOF OF CLAIM 4.26. From Lemma 4.23 the estimate (4.46) holds on the set $\mathcal{A}$. Consider now the sets $[y_{k,i}^-, y_{k,i}^+]$ and recall the definition of $y_{k,i}^*$ (in the same way we can treat the sets near the small shocks). From (4.23) and (4.25) of Lemma 4.15, we have that:

$$\left|\tilde{\gamma}(\theta, y_{k,i}^-) - \tilde{\gamma}(\theta, y_{k,i}^*-)\right| \leq L(\Delta t + \varepsilon_0 \Delta t) + C\varepsilon_0 \Delta t,$$
$$\left|\tilde{\gamma}(\theta, y_{k,i}^*+) - \tilde{\gamma}(y_{k,i}^+)\right| \leq L(\Delta t + \varepsilon_0 \Delta t) + C\varepsilon_0 \Delta t.$$

Moreover, $\tilde{\gamma}$ is defined by linear interpolations on the intervals $[y_{k,i}^-, \tilde{y}_i^\theta[$ and $]\tilde{y}_i^\theta, y_{k,i}^+]$, whose length are greater than Δt. We thus obtain the estimate

$$\left|(\tilde{\gamma}(\theta))_x\right| \leq \frac{L(\Delta t + \varepsilon_0 \Delta t) + C\varepsilon_0 \Delta t}{\Delta t} = L + (L + C)\varepsilon_0.$$

Choosing ε_0 sufficiently small the estimate (4.46) is achieved. Finally, on the set $\mathcal{K} \times [-M_0, M_0]$, the estimate is guaranteed by Lemma 4.24.
Let us now consider the estimate (4.47). For ε_0 sufficiently small, the estimate (4.34) ensures the conclusion for every θ_k. Consider first $\theta \in [\theta_k, \theta_{k+1}]$ for some $k \notin K$. From the proof of Lemma 4.23, it follows that $V(\tilde{\gamma}(\theta)) \leq V(\gamma(\theta)) + C(\varepsilon_0 + \delta_4)$. The estimate for Q has been computed in Lemma 4.21. Finally, the conclusion follows by Lemma 4.24 for $\theta \in \mathcal{K}$.
From the proof of Lemma 4.3, we have that

$$(\tilde{\gamma}(\theta_k))_x^h \exp[\beta W_h^{\tilde{\gamma}(\theta_k)}] \geq -C\delta_4$$

outside the set (4.6). Following the proof of Lemmas 4.20 and 4.23, we obtain

$$(\tilde{\gamma}(\theta))_x^h \exp[\beta W_h^{\tilde{\gamma}(\theta)}] \geq -C(\delta_4 + \varepsilon_0).$$

Hence, using (4.47), (4.48), we conclude that, for δ_4 and ε_0 sufficiently small, $\tilde{\gamma}(\theta) \in \mathcal{D}_{h,L+1/N}^*$. □

CHAPTER 5

Proof of Proposition 2.4

Let us briefly summarize what has been accomplished in the two previous sections. Given two states u^-, u^+, our aim is to construct a Lipschitz continuous semigroup of ε-solutions of (1.1) whose domain contains all suitably small BV perturbations of the Riemann data (u^-, u^+). Toward this goal, for any given $\delta_1, \delta_2, \delta > 0$ we proved that:

(i) There exists a domain $\mathcal{D}^{\delta_2}$ consisting of piecewise Lipschitz continuous functions u with $u(-\infty) = u^-$, $u(+\infty) = u^+$, $Q(u) < \varepsilon^2$. As $\delta_2 \to 0$, the domains $\mathcal{D}^{\delta_2}$ become dense on the set $\mathcal{D}$ in (2.8).

(ii) For every initial data $\bar{u} \in \mathcal{D}^{\delta_2}$, there exists a δ-accurate approximate ε-solution u of (1.1), taking values inside $\mathcal{D}^{\delta_2}$.

(iii) Let u, u' be any two δ-accurate approximations and let $\gamma_0 \colon [0,1] \mapsto \mathcal{D}^{\delta_2}$ be a Piecewise Regular Path joining $u(0)$ with $u'(0)$. Then for every $t > 0$ there exists a path $\gamma_t \colon [0,1] \mapsto \mathcal{D}^{\delta_2}$ joining $u(t)$ with $u'(t)$. The weighted length of this new path, in the metric determined by (1.39), (3.42), satisfies

$$\|\gamma_t\|_\star \leq \|\gamma_0\|_\star + C_5 t\delta \tag{5.1}$$

for some constant C_5.

We can now complete the proof of Proposition 2.4. Choose a countable dense subset

$$\mathcal{D}^* \doteq \{\bar{u}_m;\ m \geq 1\} \subset \mathcal{D} \tag{5.2}$$

of piecewise Lipschitz initial data. For every $m, \nu \geq 1$, set $\delta_1 = \delta_2 = \delta = 1/\nu$, and let $(u_{m,\nu})_{\nu \geq 1}$ be a corresponding sequence of approximate ε-solutions such that

$$u_{m,\nu} \colon [0, \infty[\mapsto \mathcal{D}^{1/\nu}, \qquad \lim_{\nu \to \infty} u_{m,\nu}(0) = \bar{u}_m. \tag{5.3}$$

Relying on a compactness argument, by possibly taking a subsequence we can assume

$$\lim_{\nu \to \infty} u_{m,\nu}(t) = u_m(t) \qquad \forall t \geq 0,\ m \geq 1, \tag{5.4}$$

for some functions $u_m \colon [0, \infty[\mapsto \mathcal{D}$.

We claim that the flow $S_t \bar{u}_m \doteq u_m(t)$ can be extended by continuity to a uniformly Lipschitz semigroup defined on the whole set $\mathcal{D}$. As in (1.48), consider the distance $d_\nu^\star$ on $\mathcal{D}^{1/\nu}$ defined by

$$d_\nu^\star \doteq \inf\Big\{ \|\gamma\|_\star\ ;\ \gamma \colon [0,1] \mapsto \mathcal{D}^{1/\nu} \text{ is a P.R.P. connecting } u \text{ with } u' \Big\}. \tag{5.5}$$

Because of the particular choice of the metric (3.42), it is clear that all these distances are uniformly equivalent to the $\mathbf{L}^1$ distance:

$$\frac{1}{C_6}\|u-u'\|_{\mathbf{L}^1} \le d^\star_\nu(u,u') \le C_6\|u-u'\|_{\mathbf{L}^1} \qquad \forall u,u' \in \mathcal{D}^{1/\nu},\ \nu \ge 1, \tag{5.6}$$

for some constant C_6 independent of ν. Observe that (5.1) implies

$$d^\star_\nu\big(u_{m,\nu}(t),\ u_{n,\nu}(t)\big) \le d^\star_\nu\big(u_{m,\nu}(0),\ u_{n,\nu}(0)\big) + C_5 t\nu^{-1}. \tag{5.7}$$

By the previous construction, for any $m,n \ge 1$ and $\varepsilon_0 > 0$, we can now choose $\nu > 1/\varepsilon_0$ so large that

$$\begin{aligned}
\big\|u_m(t) - u_n(t)\big\|_{\mathbf{L}^1} &\le \big\|u_{m,\nu}(t) - u_{n,\nu}(t)\big\|_{\mathbf{L}^1} + \varepsilon_0 \\
&\le C_6 d^\star_\nu\big(u_{m,\nu}(t),\ u_{n,\nu}(t)\big) + \varepsilon_0 \\
&\le C_6\Big\{d^\star_\nu\big(u_{m,\nu}(0),\ u_{n,\nu}(0)\big) + C_5 t\nu^{-1}\Big\} + \varepsilon_0 \\
&\le C_6\Big\{C_6\big(\big\|\bar u_m - \bar u_n\big\|_{\mathbf{L}^1} + \varepsilon_0\big) + C_5 t\varepsilon_0\Big\} + \varepsilon_0 \\
&\le L\big\|\bar u_m - \bar u_n\big\|_{\mathbf{L}^1} + C'(t+1)\varepsilon_0,
\end{aligned} \tag{5.8}$$

with $L = C_6^2$ and a suitable constant C'. Since ε_0 was arbitrary, (5.8) implies

$$\big\|S_t\bar u_m - S_t\bar u_n\big\|_{\mathbf{L}^1} \le L\|\bar u_m - \bar u_n\|\mathbf{L}^1. \tag{5.9}$$

The uniform Lipschitz continuity w.r.t. time is clear. By (5.9) the flow can thus be extended by continuity to a globally Lipschitz semigroup $S\colon \mathcal{D}\times[0,\infty[\,\mapsto \mathcal{D}$.

It remains to prove that S behaves correctly on the set of piecewise constant initial data. Let $\bar u \in \mathcal{D}$ be piecewise constant, say with jumps at the points $y_1 < \cdots < y_N$, and fix any positive time

$$\tau < \min_\alpha \frac{y_{\alpha+1} - y_\alpha}{2}.$$

For $t \in [0,\tau]$, call $u(t)$ the ε-solution of (1.1) with initial data $u(0) = \bar u$, obtained by piecing together the ε-solutions of the Riemann problems generated by the jumps of $\bar u$. We need to show that

$$S_t\bar u = u(t) \qquad \forall t \in [0,\tau]. \tag{5.10}$$

Recalling that $\mathcal{D}^*$ in (5.2) is a countable dense subset of $\mathcal{D}$, one can extract a sequence of functions $\bar u_\ell \in \mathcal{D}^*$ such that $\bar u_\ell \to \bar u$. We now have

$$S_t\bar u = \lim_{\ell\to\infty} S_t\bar u_\ell. \tag{5.11}$$

Because of (5.9), the limit in (5.11) is well defined and does not depend on the particular choice of the sequence $\bar u_\ell$.

Let $\varepsilon_0 > 0$ be given. By the previous construction, there exists a sequence of integers $\nu(\ell) \to \infty$ and a sequence of approximations $\tilde u_\ell$ with the following properties. For each $\ell \ge 1$, the function $\tilde u_\ell$ is a δ-accurate approximate ε-solution of (1.1), constructed as in the previous sections, with $\delta_1 = \delta_2 = \delta = \nu(\ell)^{-1}$. Moreover

$$\big\|\tilde u_\ell(t) - S_t\bar u_\ell\big\|_{\mathbf{L}^1} \le \varepsilon_0 \qquad t \in [0,\tau]. \tag{5.12}$$

We claim that there exists a second sequence of approximations, say $\hat{u}_\ell$, with the following properties. For each $\ell \geq 1$, the function $\hat{u}_\ell$ is a δ-accurate approximate ε-solution of (1.1), constructed as in the previous section, with $\delta_1 = \delta_2 = \delta = \nu(\ell)^{-1}$. Moreover,

$$\lim_{\ell\to\infty} \left\|\hat{u}_\ell(t) - u(t)\right\|_{\mathbf{L}^1} = 0 \qquad \forall t \in [0, \tau]. \tag{5.13}$$

The identity (5.10) is now an immediate consequence of (5.12), (5.13) and the fact that ε_0 was arbitrary. Indeed,

$$\begin{aligned}
\left\|S_t\bar{u} - u(t)\right\|_{\mathbf{L}^1} &\leq \limsup_{\ell\to\infty} \left\{ \left\|S_t\bar{u}_\ell - \tilde{u}_\ell\right\|_{\mathbf{L}^1} + \left\|\tilde{u}_\ell(t) - \hat{u}_\ell(t)\right\|_{\mathbf{L}^1} + \left\|\hat{u}_\ell(t) - u(t)\right\|_{\mathbf{L}^1} \right\} \\
&\leq \varepsilon_0 + \limsup_{\ell\to\infty} \left\|\tilde{u}_\ell(t) - \hat{u}_\ell(t)\right\|_{\mathbf{L}^1} \\
&\leq \varepsilon_0 + C_6 \cdot \limsup_{\ell\to\infty} d^\star_{\nu(\ell)}\big(\tilde{u}_\ell(t),\ \hat{u}_\ell(t)\big) \\
&\leq \varepsilon_0 + C_6 \cdot \limsup_{\ell\to\infty} \left\{ d^\star_{\nu(\ell)}\big(\tilde{u}_\ell(0),\ \hat{u}_\ell(0)\big) + C_5 t\nu(\ell)^{-1} \right\} \\
&\leq \varepsilon_0.
\end{aligned}$$

We are thus left with the task of constructing the approximations $\hat{u}_\ell$ satisfying (5.13). By an approximation argument, we can restrict the analysis to the case where each jump in $\bar{u}$ determines a single wave. More precisely, for each $\alpha = 1, \ldots, N$ there exists a family $k_\alpha \in \{1, \ldots, n\}$ such that the states $u^- = \bar{u}(y_\alpha -)$, $u^+ = \bar{u}(y_\alpha +)$ are connected by a single k_α-wave. Recalling the notation (2.2), this means

$$u^+ = \Psi^\varepsilon_{k_\alpha}(\sigma)(u^-) \tag{5.14}$$

for some wave size σ. We now consider various cases.

CASE 1: The jump at x_α is a large shock. This is the easiest case. Indeed, in a neighborhood of y_α, we can then construct approximate solutions $\hat{u}_\ell$ which coincide with u.

CASE 2: The jump at x_α is a small shock, of size $\sigma \in [-3\varepsilon, 0]$. In this case, for each ℓ, an approximate solution $\hat{u}_\ell$ is defined on a neighborhood of y_α as follows. Let the time intervals $I_{m,h}$ be as in (3.2), with $\delta_1 = \nu(\ell)^{-1}$. Recalling (3.4), define the speed of a small shock joining u^- with u^+ as

$$\lambda^\varepsilon_{k_\alpha}(u^-, u^+) \doteq (1 - n)\lambda^*_{k_\alpha} + \frac{n}{|\sigma|} \int_\sigma^0 \lambda_{k_\alpha}\big(R_{k_\alpha}(s)(u^-)\big)\, ds. \tag{5.15}$$

For each ℓ, let y_ℓ be the polygonal function such that

$$y_\ell(0) = y_\alpha, \qquad \dot{y}_\ell(t) = \begin{cases} \lambda^\varepsilon_{k_\alpha}(u^-, u^+) & \text{if } t \in I_{m,k_\alpha} \text{ for some } m, \\ \lambda^*_{k_\alpha} & \text{otherwise.} \end{cases} \tag{5.16}$$

Choosing a sequence $\varepsilon_\ell \to 0$ sufficiently fast, we can now define the approximate ε-solutions $\hat{u}_\ell$ by setting

(5.17)
$$\hat{u}_\ell(t,x) = \begin{cases} u^- & \text{if } x \le y(t), \\ u^+ & \text{if } x \ge y(t) + \varepsilon_\ell, \\ u^+ & \text{if } x \in \big[y(t),\ y(t)+\varepsilon_\ell\big],\ t \in I_{m,k_\alpha} \text{ for some } m, \\ R_{k_\alpha}\left(\frac{x-y(t)}{\varepsilon_\ell}\right)(u^-) & \text{if } x \in \big[y(t),\ y(t)+\varepsilon_\ell\big],\ t \notin \bigcup_{m\ge 0} I_{m,k_\alpha}. \end{cases}$$

We now observe that, in a neighborhood of y_α, the ε-solution u satisfies

$$u(t,x) = \begin{cases} u^- & \text{if } x < y_\alpha + t\lambda^\varepsilon_{k_\alpha}(u^-,u^+), \\ u^+ & \text{if } x > y_\alpha + t\lambda^\varepsilon_{k_\alpha}(u^-,u^+). \end{cases} \tag{5.18}$$

As $\ell \to \infty$ and $\delta_2 = \nu(\ell)^{-1} \to 0$, the sequence $y_\ell(t)$ converges uniformly to $t\lambda^\varepsilon_{k_\alpha}(u^-,u^+)$. Comparing (5.17) with (5.18), the convergence $\hat{u}_\ell \to u$ is thus clear.

CASE 3: The jump at y_α is a rarefaction wave, so that (5.14) holds for some $\sigma > 0$. In this case, we choose a sequence $\varepsilon_\ell \to 0$ sufficiently fast and define the functions $\hat{u}_\ell$ in a neighborhood of y_α as follows. Each $\hat{u}_\ell$ is the Lipschitz continuous solution of the quasilinear hyperbolic system

$$u_t + A(t,u)u_x = 0, \tag{5.19}$$

with initial condition

$$\hat{u}_\ell(0,x) = \begin{cases} u^- & \text{if } x < y_\alpha, \\ u^+ & \text{if } x > y_\alpha + \varepsilon_\ell, \\ R_{k_\alpha}\big((x-y_\alpha)/\varepsilon_\ell\big)(u^-) & \text{if } x \in [y_\alpha,\ y_\alpha + \varepsilon_\ell]. \end{cases} \tag{5.20}$$

Here $A(t,u)$ is the matrix having the same eigenvectors $r_1(u), \ldots, r_n(u)$ as $A(u) = DF(u)$, but whose eigenvalues are

$$\lambda^*_1,\ \ldots,\ \lambda^*_{h-1},\ \lambda^*_h + n\big(\lambda_h(u) - \lambda^*_h\big),\ \ldots,\lambda^*_n \qquad \text{if } t \in I_{m,h}.$$

The time intervals $I_{m,h}$ are as in (3.2), with $\delta_1 = \nu(\ell)^{-1}$. It is now clear that all functions $\hat{u}_\ell$ remain Lipschitz continuous in a neighborhood of y_α, because they only contain rarefaction waves. Moreover, the convergence $\hat{u}_\ell \to u$ holds. This completes the proof of Proposition 2.4.

CHAPTER 6

Proof of Proposition 2.5

In this section we construct the semigroup S^ε generated by the ε-approximate Riemann Solver, and prove Proposition 2.5. This will accomplish Step 2 toward the proof of Theorem 2.3.

Let $\bar{u}$ be an initial data with compact support and suitably small total variation. Using the wave-front tracking algorithm described in Section 2, we construct a sequence of piecewise constant approximate solutions u_ν, with $u_\nu(0) \to \bar{u}$, such that

(i) The total variation of $u_\nu(t,\cdot)$ remains uniformly small,
(ii) The maximum size of the rarefaction fronts in u_ν approaches zero,
(iii) The total strength of all non-physical waves in u_ν approaches zero.

By possibly taking a subsequence, we can assume that $u_\nu \to u$ in $\mathbf{L}^1_{loc}$. We claim that the limit function u is unique, and provides a viscosity ε-solution to the corresponding Cauchy problem (1.1)-(1.2).

To prove uniqueness, let u_ν, w_ν, $\nu \geq 1$, be sequences of approximate solutions constructed by wave-front tracking, and assume that

$$\lim_{\nu\to\infty} u_\nu(0,\cdot) = \lim_{\nu\to\infty} w_\nu(0,\cdot) = \bar{u},$$

while $u_\nu \to u$, $w_\nu \to w$ in $\mathbf{L}^1_{loc}$. Since both u and w are continuous as maps from $[0,\infty[$ into $\mathbf{L}^1$, if $u \neq w$ there exists a largest time τ such that $u(t) = w(t)$ for all $t \in [0,\tau]$. Fix any $\bar{x} \in \mathbb{R}$. We will prove that $u(t,x) = w(t,x)$ a.e. in a region of the form

$$\Gamma \doteq \Big\{(t,x);\ t \in [\tau,\ \tau+\delta],\ |x-\bar{x}| \leq \rho - (t-\tau)\Big\} \tag{6.1}$$

for some $\rho, \delta > 0$. By possibly choosing subsequences, we can assume the weak convergence

$$\text{Tot.Var.} u_\nu(\tau,\cdot) \rightharpoonup \mu_1, \qquad \text{Tot.Var.} w_\nu(\tau,\cdot) \rightharpoonup \mu_2,$$

for some positive measures μ_1, μ_2. Choose $\rho > 0$ small enough so that

$$\mu_i\Big([\bar{x}-\rho,\ \bar{x}+\rho] \setminus \{\bar{x}\}\Big) \leq \eta^3 \qquad\qquad i = 1,2. \tag{6.2}$$

Here $\eta \ll 1$ is the constant in (2.8). For $t \geq \tau$, consider the interval

$$J(t) \doteq \big[\bar{x} - \rho + (t-\tau),\ \bar{x} + \rho - (t-\tau)\big]. \tag{6.3}$$

By [**B5**], for each $t > \tau$ the restriction of the interaction potential of u_ν, w_ν to $J(t)$ satisfies

$$Q\big(u_\nu\big|_{J(t)}\big) \leq 2\eta^3, \qquad Q\big(w_\nu\big|_{J(t)}\big) \leq 2\eta^3, \tag{6.4}$$

for all ν suitably large (depending on t). Therefore, calling $u^- = u(\tau,\bar{x}-)$, $u^+ = u(\tau,\bar{x}+)$ we can find a sequence of times t_m decreasing to τ and integers $\nu(m)$ such that all functions $\tilde{u}_\nu(t_m,\cdot), \tilde{w}_\nu(t_m,\cdot)$ with $\nu \geq \nu(m)$ lie in the domain $\mathcal{D}_{(u^-,u^+)}$ of the semigroup S constructed in Proposition 2.4. Here we define

$$\tilde{u}_\nu(t_m,x) \doteq \begin{cases} u_\nu(t_m,x) & \text{if} \quad x \in J(t_m), \\ u^- & \text{if } x < \bar{x} - \rho + (t_m - \tau), \\ u^+ & \text{if } x > \bar{x} + \rho - (t_m - \tau), \end{cases}$$

and similarly for $\tilde{w}_\nu(t_m,\cdot)$. For $t \geq \tau$ we also define

$$\tilde{u}(t,x) \doteq \begin{cases} u(t,x) & \text{if } x \in J(t), \\ u^- & \text{if } x < \bar{x} - \rho + (t - \tau), \\ u^+ & \text{if } x > \bar{x} + \rho - (t - \tau), \end{cases} \tag{6.5}$$

and similarly for $\tilde{w}$. We now observe that, on any region of the form

$$\Gamma_m \doteq \Big\{(t,x);\ t \in [t_m,\ \tau+\delta],\ x \in J(t)\Big\},$$

the functions u, w are limits of wave-front tracking approximations taking values within the domain of the Lipschitz semigroup S. The same arguments used in Theorem 1.1 of [**B5**] thus imply

$$u(t,x) = \big(S_{t-t_m}\tilde{u}(t_m)\big)(x), \qquad w(t,x) = \big(S_{t-t_m}\tilde{w}(t_m)\big)(x) \qquad \text{for } (t,x) \in \Gamma_m.$$

Calling L the Lipschitz constant of the semigroup S, for $t > t_m$ we now have the estimate

$$\begin{aligned} \int_{J(t)} \big|u(t,x) - w(t,x)\big|\,dx &\leq \Big\|S_{t-t_m}\tilde{u}(t_m) - S_{t-t_m}\tilde{w}(t_m)\Big\|_{\mathbf{L}^1} \\ &\leq L\big\|\tilde{u}(t_m) - \tilde{w}(t_m)\big\|_{\mathbf{L}^1} \\ &\leq L\Big(\big\|\tilde{u}(t_m) - \tilde{u}(\tau)\big\|_{\mathbf{L}^1} + \big\|\tilde{u}(\tau) - \tilde{w}(\tau)\big\|_{\mathbf{L}^1} + \big\|\tilde{w}(\tau) - \tilde{w}(t_m)\big\|_{\mathbf{L}^1}\Big). \end{aligned} \tag{6.6}$$

As $m \to \infty$, for each fixed $t > \tau$ the right hand side of (6.6) approaches zero. Hence $u = w$ on Γ.

By the boundedness of the supports of u, w we can now choose a constant R large enough so that $u(t,x) = w(t,x) = 0$ for $t \in [\tau,\ \tau+1]$, $|x| > R$. Moreover, for every point $(\tau, \bar{x})$ with $|x| \leq R$ there exists a set Γ of the form (6.1) on which $u = w$. By a compactness argument it follows that $u = w$ on a strip of the form $[\tau,\ \tau+\delta_0] \times \mathbb{R}$. This contradicts the maximality of τ, proving that the limit solution u obtained by wave-front tracking is unique. From now on, this limit will be indicated by the semigroup notation

$$u(t,\cdot) = S_t^\varepsilon \bar{u}. \tag{6.7}$$

The continuity of the map S_t^ε w.r.t. $\bar{u}$ is an immediate consequence of uniqueness. Indeed, consider a sequence of initial conditions with sufficiently small total variation $\bar{u}_\nu \to \bar{u}$. Given $T > 0$, for each $\nu \geq 1$ there exists an approximate solution u_ν obtained by wave-front tracking such that

$$\big\|u_\nu(0,\cdot) - \bar{u}_\nu\big\|_{\mathbf{L}^1} \leq \frac{1}{\nu}, \qquad \big\|u_\nu(t,\cdot) - S_t^\varepsilon \bar{u}_\nu\big\|_{\mathbf{L}^1} \leq \frac{1}{\nu} \qquad t \in [0,T].$$

By uniqueness it now follows

$$S_t^\varepsilon \bar{u} = \lim_{\nu\to\infty} u_\nu(t,\cdot) = \lim_{\nu\to\infty} S_t^\varepsilon \bar{u}_\nu \qquad\qquad t \in [0,T],$$

proving the continuity of the map S_t^ε.

To show that u is a viscosity ε-solution, we observe that, for any point $(\tau, \bar{x})$, for $\rho > 0$ small enough, the truncated function $\tilde{u}$ defined at (6.5) lies within the domain $\mathcal{D}_{(u^-,u^+)}$ of one of the local semigroups S constructed in Step 1. By uniqueness and finite propagation speed, it follows that

$$u(t,x) = \big(S_{t-\tau}\tilde{u}(\tau)\big)(x) \tag{6.8}$$

for all (t,x) with $t \geq \tau$, $|x - \bar{x}| < \rho - (t - \tau)$. Since the right hand side of (6.8) is a viscosity ε-solution of (1.1), the function u satisfies the estimates (1.13), (1.14) at the point $(\tau, \bar{x})$. This completes the proof of Proposition 2.5.

CHAPTER 7

Proof of Proposition 2.7

Aim of this section is to show that, if $\tilde{u}$ is a structurally stable ε-solution, then all suitably accurate approximations constructed by our algorithm will have the same wave-front structure as $\tilde{u}$. The first step in the proof is to establish a decay estimate similar to (1.21), valid for the piecewise Lipschitz approximate solutions constructed in Sections 3-4. We start with a simple estimate for an impulsive O.D.E., based on a comparison argument.

LEMMA 7.1. *Let b, g be non-negative integrable functions on the interval $[\tau, t]$ and let $w\colon [\tau, t] \mapsto [0, \infty[$ satisfy the impulsive differential inequality*

$$\dot{w}(s) \leq b(s)w(s) - g(s), \quad w(\tau) = w_0, \quad w(\tau_i+) \leq b_i \cdot w(\tau_i-) \qquad i = 1, \dots, N,$$

with jumps at times $\tau_i \in [\tau, t]$. Assume that $w_0 \geq 0$ and $b_i \geq 1$ for all i. Then

$$w(t) \leq \exp\left\{\int_\tau^t b(s)\, ds\right\} \cdot \left(\prod_{i=1}^N b_i\right) w_0 - \int_\tau^t g(s)\, ds. \tag{7.1}$$

Indeed, calling z the solution to the impulsive differential equation

$$\dot{z}(s) = b(s)z(s) - g(s), \qquad z(\tau) = w_0, \qquad z(\tau_i+) = b_i \cdot z(\tau_i-) \qquad (i = 1, \dots, N),$$

a comparison argument yields

$$\begin{aligned} w(t) \leq z(t) = &\exp\left\{\int_\tau^t b(s)\, ds\right\} \cdot \left(\prod_{i=1}^N b_i\right) w_0 \\ &- \int_\tau^t \left(\prod_{\tau_i \in [s,t]} b_i\right) \exp\left\{\int_s^t b(r)\, dr\right\} \cdot g(s)\, ds. \end{aligned}$$

Since $b_i \geq 1$ for all i, this implies (7.1).

Now consider a piecewise Lipschitz approximate solution $u = u(t,x)$ constructed as in Sections 3-4. More precisely, given a time step δ_1 and a shock layer width δ_2, we consider an approximate solution $u = u(t,x)$ satisfying the following.
- Each $u(t, \cdot)$ is piecewise Lipschitz, with large shocks (of strength $|\sigma_\alpha| > \varepsilon^9$) at points y_α, $\alpha \in \mathcal{S}$, plus other small shocks (of strength $|\sigma_\beta| < 3\varepsilon$) at points y_β, $\beta \in \mathcal{S}'$.
- For $t \in I_{m,h}$, $m \geq 0$, $h \in \{1, \dots, n\}$ as in (3.2), u provides a solution to a quasi-linear system of the form (3.8), where $A^{(h)}$ is a matrix with the same eigenvectors as $A(u)$. Concerning the eigenvalues $\lambda_i^{(h)}$ of $A^{(h)}$, the following holds. All characteristic speeds $\lambda_i^{(h)}$, $i \neq h$ are locally constant, with jumps only along the big

shocks. The h-characteristic speed $\lambda_h^{(h)}$ is constant inside the shock layers $[y_\alpha^*,\ y_\alpha[$, $]y_\alpha,\ y_\alpha^{**}]$ around each big shock, and is genuinely nonlinear outside, so that

$$\nabla\lambda_h^{(h)} \cdot r_h > n\kappa' > 0 \qquad\qquad x \notin [y_\alpha^*,\ y_\alpha^{**}] \doteq [y_\alpha - \delta_2,\ y_\alpha + \delta_2]. \tag{7.2}$$

- If u has a shock in the k_α-th family at y_α, the wave speed $\lambda_{k_\alpha}^{(h)}$ across the shock layer satisfies the estimates

$$\min\left\{\left|\lambda_{k_\alpha}^{(h)}(y_\alpha+) - \dot y_\alpha\right|, \left|\lambda_{k_\alpha}^{(h)}(y_\alpha-) - \dot y_\alpha\right|\right\} > \kappa''|\sigma_\alpha| \tag{7.3}$$

for some constant $\kappa'' > 0$. Moreover, if $k_\alpha = h$ one has

$$\max\left\{\left|\lambda_h^{(h)}(y_\alpha^*+) - \lambda_h^{(h)}(y_\alpha^*-)\right|, \left|\lambda_h^{(h)}(y_\alpha^{**}+) - \lambda_h^{(h)}(y_\alpha^{**}-)\right|\right\} \le C|\sigma_\alpha| \tag{7.4}$$

- Restartings are performed at due times, according to the procedures described in Section 4, with a suitable degree of accuracy which will be made precise at a later stage of the analysis.

Let there be given a time interval $[\tau, \bar t]$, an index $h \in \{1, \dots, n\}$ and $\eta > 0$. Our aim is to estimate, at time $t = \bar t$, the amount of positive h-waves in u with density $> \eta$, i.e. the quantity

$$V_h^{\eta+} = \sum_{\sigma_\alpha^h > 0} \sigma_\alpha^h + \int_{\{u_x^h > \eta\}} u_x^h(\bar t, x)\, dx. \tag{7.5}$$

We here assume that $u(\bar t, \cdot)$ has jumps at points y_α and call $\sigma_\alpha^1, \dots, \sigma_\alpha^n$ the waves in the Riemann problem determined by the jump at y_α.

The only contributions to the first summation in (7.5) are due to h-rarefaction fronts produced by the interaction of two large shocks at some time $t \in [\bar t - \delta_1,\ \bar t]$. This summation can thus be estimated in terms of the local amount of interaction, namely

$$\sum_{\sigma_\alpha^h > 0} \sigma_\alpha^h \le C\big[Q(\bar t - \delta_1) - Q(\bar t)\big] \tag{7.6}$$

for some constant C. To estimate the integral term in (7.5), we study the evolution of the gradient component u_x^h along h-characteristics, for $t \in [\tau, \bar t]$. For any fixed x, call $t \mapsto y^x(t)$ the h-characteristic line through the point $(\bar t, x)$. Recalling (3.2) and (3.4)–(3.7), y^x is thus defined as the solution to the backward Cauchy problem

$$y^x(\bar t) = x, \qquad \dot y^x(t) = \lambda_h^{(j)}\big(t, y^x(t), u\big) \qquad \text{if} \quad t \in I_{m,j} \quad \text{for some } m, j. \tag{7.7}$$

We distinguish two cases:

(a) The characteristic y^x is defined for all $t \in [\tau, \bar t]$ and u_x^h remains positive and uniformly bounded along y^x.

(b) Either $u_x^h(t_0, y^x(t_0)) \le 0$ at some time $t_0 \in [\tau, \bar t]$, or else the characteristic y^x originates from a centered rarefaction fan, generated by the interaction of two shocks.

We shall consider separately the sets J_a, J_b of points x for which the alternative (a) or (b) holds. Assume first $x \in J_a$. Consider the scalar functions

$$v(t) \doteq \frac{\partial y^x(t)}{\partial x}, \tag{7.8}$$

$$z(t) \doteq u_x^h\big(t, y^x(t)\big) \cdot v(t). \tag{7.9}$$

Observe that, for Δx small, the quantity $z(t)\cdot\Delta x$ roughly determines the amount of h-waves contained in the infinitesimal segment $\big[y^x(t),\ y^{x+\Delta x}(t)\big]$. The scalar quantities z, u_x^h and v evolve continuously along the characteristic y^x, except for a finite number of times τ_i at which they experience a discontinuity. These discontinuities occur when the characteristic y^x crosses a shock or a shock layer, and when a restarting procedure is applied. Let us first describe the smooth evolution equations and then treat the impulses due to discontinuities. On time intervals where it is continuous, the function v satisfies the equation

$$\dot v(t) = \big(\lambda_h^{(j)}(t, y^x(t), u)\big)_x \cdot v(t) \qquad t\in I_{m,j}. \tag{7.10}$$

Similarly, on time intervals where it is continuous, the gradient component u_x^h satisfies an equation of the form (1.18). Hence, for $t\in I_{m,j}$ we have

$$\begin{aligned}
\dot z(t) &= \frac{du_x^h}{dt}\cdot v + u_x^h\cdot\frac{dv}{dt}\\
&= \big[(u_x^h)_t + (u_x^h)_x\,\lambda_h^{(j)}(u)\big]\cdot v + u_x^h\cdot(\lambda_h^{(j)})_x\,v =\\
&= \left[-(\lambda_h^{(j)})_x\,u_x^h + \sum_{i\neq k}\tilde G_{hik}(u)\,u_x^i\,u_x^k\right]\cdot v + u_x^h\cdot(\lambda_h^{(j)})_x\,v =\\
&= \left[\sum_{i\neq k}\tilde G_{hik}(u)\,u_x^i\,u_x^k\right]\cdot v
\end{aligned} \tag{7.11}$$

for some functions $\tilde G_{hik}$ whose expression is easily derived from (1.18).

When $t\in I_{m,j}$, from (7.10) it follows

$$\dot v(t) = \sum_{i=1}^{n}(\nabla\lambda_h^{(j)}\cdot r_i)u_x^i\cdot v(t) \ge a(t)z(t) - q(t)v(t), \tag{7.12}$$

where

$$a(t) \doteq (\nabla\lambda_h^{(j)}\cdot r_h)(t, y^x, u), \qquad q(t)\doteq C\sum_{i\neq h}\Big|u_x^i\,(t, y^x(t))\Big|. \tag{7.13}$$

Notice that $a(t)=0$ whenever $t\notin\bigcup_m I_{m,h}$. Recalling (7.2) we obtain

$$\int_\tau^{\bar t} a(t)\,dt \ge \kappa'(\bar t-\tau) - C(\delta_1+\delta_2) \ge \frac{\kappa'}{2}(\bar t-\tau), \tag{7.14}$$

provided that $\delta_1,\delta_2>0$ are chosen so that

$$\delta_1+\delta_2 \le \frac{\kappa'(\bar t-\tau)}{2C}.$$

Indeed, the characteristic $y^x(\cdot)$ will spend a time $O(\delta_2)$ inside the shock layers. Moreover,

$$\left|\frac{\bar t-\tau}{n} - \text{meas}\left([\tau,\ \bar t]\cap\Big(\bigcup_{m\ge 0} I_{m,h}\Big)\right)\right| \le \frac{2\delta_1}{n}.$$

Concerning the impulses, observe that the function v is continuous at restarting times and has jumps at times where the characteristic y^x crosses either a large shock or a shock layer. In all such cases, denoting by y_α the location of the large shock, one has

$$v(t+) = \frac{\dot y^x(t+) - \dot y_\alpha(t)}{\dot y^x(t-) - \dot y_\alpha(t)} \cdot v(t-).$$

Therefore, at a crossing time τ_i we have the estimate

$$v(\tau_i-) \le b_i v(\tau_i+), \qquad b_i \doteq \left| \frac{\dot y^x(\tau_i+) - \dot y^x(\tau_i-)}{\dot y^x(\tau_i-) - \dot y_\alpha(\tau_i)} \right|. \tag{7.15}$$

We now observe that the integral of q in (7.13) is controlled by the amount of waves that cross the line y^x. Moreover, y^x can cross at most one shock layer of a shock y_α with $k_\alpha = h$, because afterwards this h-characteristic will impinge on the shock. For all other shocks, by strict hyperbolicity we have

$$\big|\dot y^x(\tau_i-) - \dot y_\alpha(\tau_i)\big| \ge \Delta\lambda > 0.$$

Recalling (7.3)-(7.4) we thus have

$$\sum_i |b_i - 1| = O(1).$$

The previous arguments yield an estimate of the form

$$\int_\tau^{\bar t} q(t)\,dt \le C, \qquad \prod_i b_i \le \exp\left\{\sum_i |b_i - 1|\right\} \le C,$$

$$\exp\left\{\int_\tau^{\bar t} q(t)\,dt\right\} \cdot \left(\prod_i b_i\right) \le C_1 \tag{7.16}$$

for some constant C_1.

We now apply Lemma 7.1 to the function $w(t) \doteq v(-t)$ on the time interval $[-\bar t, -\tau]$. By (7.2) it follows

$$w(-\tau) \le \exp\left\{\int_\tau^{\bar t} q(t)\,dt\right\} \left(\prod_i b_i\right) w(-\bar t) - \int_\tau^{\bar t} a(t) z(t)\,dt$$

Since $w(-\bar t) = v(\bar t) = 1$, by (7.16) this yields

$$0 \le v(\tau) \le C_1 - \int_\tau^{\bar t} a(t) z(t)\,dt. \tag{7.17}$$

Next, consider the evolution equation for z, along the h-characteristic y^x. Define the rate of interaction along y^x, i.e.

$$f(t) \doteq \sum_{i<k} \big[\lambda_k^{(j)} - \lambda_i^{(j)}\big] |u_x^i| |u_x^k| \qquad t \in I_{m,j}, \tag{7.18}$$

where the right hand side of (7.18) is computed at the point $\big(t, y^x(t)\big)$. By strict hyperbolicity, the differences $\lambda_k - \lambda_i$ are uniformly positive. Recalling (7.11), on time intervals where z is continuous there holds

$$\big|\dot z(t)\big| \le C f(t) \cdot v(t). \tag{7.19}$$

Concerning the jumps in z, at a time τ_i where y^x crosses a big shock, say at y_α, we have the estimate

$$|z(\tau_i+) - z(\tau_i-)| \leq C\Lambda_\alpha(\tau_i) \cdot v(\tau_i-), \tag{7.20}$$

where Λ_α is the instantaneous rate of interaction along y_α, defined as in (3.35). From (7.17) and (7.19) we deduce

$$\begin{aligned} 0 &\leq C_1 - \int_\tau^{\bar t} a(t)\left(z(\bar t) - C\int_t^{\bar t} f(s)v(s)\,ds - \sum_i |z(\tau_i+) - z(\tau_i-)|\right) dt \\ &\leq C_1 - z(\bar t)\int_\tau^{\bar t} a(t)\,dt + C_2(\bar t - \tau)\left(\int_\tau^{\bar t} f(t)v(t)\,dt + \sum_i |z(\tau_i+) - z(\tau_i-)|\right) \end{aligned} \tag{7.21}$$

for some constant C_2. Dividing by $\int a(t)$ and recalling (7.14) one obtains

$$z(\bar t) \leq \frac{1}{\kappa(\bar t - \tau)} + C_3\left(\int_\tau^{\bar t} f(t)v(t)\,dt + \sum_i |z(\tau_i+) - z(\tau_i-)|\right) \tag{7.22}$$

for some constants $C_3, \kappa > 0$. We now let x vary inside the set J_a, calling v^x, z^x, f^x the corresponding functions along the characteristic y^x. Since $z^x(\bar t) > \eta$ on J_a, from (7.22) it follows

$$\begin{aligned} \left(1 - \frac{1}{\eta\kappa(\bar t - \tau)}\right)\int_{J_a} z^x(\bar t)\,dx &\leq \int_{J_a}\left(1 - \frac{1}{z^x(\bar t)\kappa(\bar t - \tau)}\right) z^x(\bar t)\,dx \\ &\leq \int_{J_a} C_3\left(\int_\tau^{\bar t} f^x(t)v^x(t)\,dt + \sum_i |z^x(\tau_i^x+) - z^x(\tau_i^x-)|\right) dx. \end{aligned} \tag{7.23}$$

Since $v(t)$ is the Jacobian of the transformation $x \mapsto y^x(t)$, by (7.18) the double integral of $f \cdot v$ is controlled in terms of the total amount of interaction, i.e.

$$\int_{\mathbb{R}}\int_\tau^{\bar t} f^x(s)v^x(s)\,ds dx \leq C[Q(\tau) - Q(\bar t)]. \tag{7.24}$$

Moreover, recalling (7.20) and estimating separately the sum of jumps in z occurring at crossings of shocks and at restarting times, for any given $\varepsilon_0 > 0$ we have the estimates

$$\int_{\mathbb{R}}\left(\sum_{\tau_i \in \text{Cross.}} |z^x(\tau_i^x+) - z^x(\tau_i^x-)|\right) dx \leq C[Q(\tau) - Q(\bar t)]. \tag{7.25}$$

$$\int_{\mathbb{R}}\left(\sum_{\tau_i \in \text{Rest.}} |z^x(\tau_i^x+) - z^x(\tau_i^x-)|\right) dx \leq \varepsilon_0, \tag{7.26}$$

provided that the restarting procedures are suitably accurate.
Since $z^x(\bar t) = u_x^h(\bar t, x)$, the estimates (7.23)–(7.26) imply

$$\int_{J_a} u_x^h(\bar t, x)\,dx \leq C_4[Q(\tau) - Q(\bar t) + \varepsilon_0]\left(1 - \frac{1}{\eta\kappa(\bar t - \tau)}\right)^{-1}. \tag{7.27}$$

for some constant C_4.

Finally, we consider the integral of u_x^h over J_b. Observe that, if $x \in J_b$, there exists a time $\tau_0 \in [\tau, \bar{t}]$ such that $v^x(\tau_0) = 0$ or $z^x(\tau_0) \leq 0$. Therefore,

$$z^x(\bar{t}) \leq C_3 \left(\int_\tau^{\bar{t}} f^x(t) v^x(t)\, dt + \sum_i \left| z^x(\tau_i+) - z^x(\tau_i-) \right| \right). \tag{7.28}$$

Integrating over J_b and using (7.24)–(7.26) we obtain

$$\begin{aligned} \int_{J_b} u_x^h(\bar{t}, x)\, dx &\leq C_3 \int_{J_b} \left(\int_\tau^{\bar{t}} f^x(t) v^x(t)\, dt + \sum_i \left| z^x(\tau_i+) - z^x(\tau_i-) \right| \right) dx \\ &\leq C_4 \big[Q(\tau) - Q(\bar{t}) + \varepsilon_0 \big]. \end{aligned} \tag{7.29}$$

where $\varepsilon_0 > 0$ can be taken arbitrarily small by increasing the accuracy of the restarting procedures. The bounds (7.6), (7.27) and (7.29) together yield

$$V_h^{\eta+} \leq C\big[Q(\tau) - Q(\bar{t}) + \varepsilon_0\big] \left(1 - \frac{1}{\eta\kappa(\bar{t} - \tau)} \right)^{-1}. \tag{7.30}$$

Toward the proof of Proposition 2.7, we establish another lemma. Roughly speaking it shows that, if the $\mathbf{L}^1$ distance between u, $\tilde{u}$ is small and if the amount of steep positive waves in u can be estimated, then the total amount of waves in u can be bounded in terms of the total amount of waves in $\tilde{u}$.

In the following we consider two intervals $J = [a, b]$, $J' = [a - \delta_0,\ b + \delta_0]$. By $V(u; J)$ we denote the amount of waves in u inside the interval J. The total amount of positive waves of density $> \eta$ is written $V^{\eta+}(u; J)$.

LEMMA 7.2. *Consider two functions $u, \tilde{u} \colon J' \mapsto \Omega$. For some constant C, the following holds. Assume that*

$$\int_{J'} \left| u(x) - \tilde{u}(x) \right|\, dx \leq \delta_1, \qquad V(\tilde{u}; J') \leq \delta_2, \qquad V^{\eta+}(u, J') \leq \delta_3. \tag{7.31}$$

Then the total amount of waves of u on J is bounded by

$$V(u; J) \leq C \left[\delta_2 + \eta(b - a + \delta_0) + \delta_3 + \frac{\delta_1}{\delta_0} \right]. \tag{7.32}$$

PROOF. We can assume that there exists $c_0 > 0$ and a unit vector $\mathbf{e}$ such that

$$c_0 \leq \langle \mathbf{e},\ r_i(u) \rangle \leq 1 \qquad i = 1, \dots, n,\ u \in \Omega.$$

$$\sigma \leq \langle \mathbf{e},\ S_i(\sigma)(u) - u \rangle \leq c_0 \sigma \qquad i = 1, \dots, n,\ \sigma < 0,\ u \in \Omega.$$

Here $S_i(\sigma)(u)$ is the state connected to u by an i-shock of size $\sigma < 0$. The total amount of positive waves of u on $[a, b]$ is

$$V^+(u; J) \leq \delta_3 + n\eta(b - a).$$

Call $V^-(u; J)$ the amount of negative waves of u on $[a, b]$. Then

$$\langle \mathbf{e},\ u(b) - u(a) \rangle \leq V^+(u; J) - c_0 V^-(u; J).$$

On the other hand,

$$\langle \mathbf{e},\ \tilde{u}(b) - \tilde{u}(a) \rangle \geq -V(\tilde{u}; J).$$

Define

$$\alpha_1 \doteq \langle \mathbf{e},\ u(a) - \tilde{u}(a)\rangle, \qquad \alpha_2 \doteq \langle \mathbf{e},\ \tilde{u}(b) - u(b)\rangle.$$

Observe that

$$\begin{aligned} \alpha_1 + \alpha_2 &= \langle \mathbf{e},\ u(a) - u(b)\rangle + \langle \mathbf{e},\ \tilde{u}(b) - \tilde{u}(a)\rangle \\ &\geq c_0 V^-(u;J) - V^+(u;J) - V(\tilde{u};J). \end{aligned} \tag{7.33}$$

If $\alpha_1 + \alpha_2 \leq 0$, from (7.3) it follows

$$V^-(u,J) \leq \frac{1}{c_0}\Big(V^+(u;J) - V(\tilde{u};J)\Big) \leq \frac{1}{c_0}\big[\delta_3 + n\eta(b-a) + \delta_2\big]. \tag{7.34}$$

If $\alpha_1 + \alpha_2 > 0$, to fix the ideas assume $\alpha_2 \geq \alpha_1$, the other case being entirely similar. For $x \in [b,\ b+\delta_0]$ we have

$$\begin{aligned} \langle \mathbf{e},\ \tilde{u}(x) - u(x)\rangle \geq& \alpha_2 - V\big(\tilde{u};[b,x]\big) - V^+\big(u;[b,x]\big) \\ \geq& \frac{1}{2}\Big(c_0 V^-(u;J) - V^+(u,J) - V(\tilde{u};J)\Big) \\ &- V\big(\tilde{u};[b,x]\big) - V^+\big(u;[b,x]\big) \\ \geq& \frac{c_0 V^-(u;J)}{2} - V^+\big(u;[a,x]\big) - V\big(\tilde{u};[a,x]\big) \\ \geq& \frac{c_0 V^-(u;J)}{2} - \big\{\delta_3 + n\eta(b-a+\delta_0)\big\} - V(\tilde{u};J'). \end{aligned}$$

Therefore,

$$\begin{aligned} \delta_1 &\geq \int_b^{b+\delta_0} \big|u(x) - \tilde{u}(x)\big|\,dx \geq \int_b^{b+\delta_0} \langle \mathbf{e},\ \tilde{u}(x) - u(x)\rangle\,dx \\ &\geq \delta_0\left[\frac{c_0 V^-(u;J)}{2} - V(\tilde{u};J') - \delta_3 - n\eta(b-a+\delta_0)\right]. \end{aligned}$$

This yields the bound

$$V^-(u;J) \leq \frac{2}{c_0}\left[\frac{\delta_1}{\delta_0} + \delta_2 + \delta_3 + n\eta(b-a+\delta_0)\right]. \tag{7.35}$$

The bounds (7.34) and (7.35) clearly imply (7.32) for some constant C.

Proof of Proposition 2.7. For simplicity, we shall assume that our approximate solution satisfies the decay estimate (1.21). The same arguments can be easily adapted to the case where (7.30) holds, provided that $\varepsilon_0 > 0$ is sufficiently small.

Consider CASE 1 first. Define the intervals

$$J'(t) \doteq \big[\bar{x} - 9(\tau - t),\ \bar{x} + 9(\tau - t)\big], \qquad J(t) \doteq \big[\bar{x} - 8(\tau - t),\ \bar{x} + 8(\tau - t)\big]. \tag{7.36}$$

By assumption (see **[B-LF2]**), there exists $r_0 > 0$ such that

$$V\big(\tilde{u}(t);\ J'(t)\big) \leq 2\varepsilon^2 \qquad \forall t \in \ [\tau - r_0,\ \tau[\,. \tag{7.37}$$

By (1.8), $Q(u) \leq C_0$ for all all functions u under consideration. Let N be an integer such that $N\varepsilon^2 > C_0$. Define

$$r_m \doteq (\varepsilon^2\kappa)^m r_0 \qquad t_m \doteq \tau - r_m, \qquad m = 1, \ldots, N, \tag{7.38}$$

where κ is the constant in (1.21). We can assume that $\varepsilon^2\kappa < 1/2$, so that $r_{m-1} - r_m > r_{m-1}/2$. Consider any solution $u \in \mathcal{D}$ suitably close to $\tilde{u}$, so that

$$\int_{J'(t_m)} \big|u(t_m,x) - \tilde{u}(t_m,x)\big|\, dx < \varepsilon^2 r_m \qquad m = 1,\ldots,N. \tag{7.39}$$

Since the function $t \mapsto Q(t) \doteq Q\big(u(t)\big)$ is non-negative and $Q\big(u(t_0)\big) < C_0$, there exists some integer $m \leq N$ such that

$$Q(t_{m-1}) - Q(t_m) < \varepsilon^2. \tag{7.40}$$

Choosing $\eta = 4/\kappa r_{m-1}$, the decay estimate (1.21) yields

$$V^{\eta+}\big(u(t_m)\big) \leq C_3\big[Q(t_{m-1}) - Q(t_m)\big]\left(1 - \frac{1}{\eta\kappa(t_m - t_{m-1})}\right)^{-1} \leq 2C_3\varepsilon^2. \tag{7.41}$$

Indeed, $t_m - t_{m-1} = r_{m-1} - r_m > r_{m-1}/2$. Applying Lemma 7.2 to the intervals $J(t_m)$, $J'(t_m)$, again with $\eta = 4/\kappa r_{m-1}$, and using (7.37)–(7.39) and (7.41), we find

$$\begin{aligned} V\big(u(t_m); J(t_m)\big) &\leq C\left[V\big(\tilde{u}(t_m); J'(t_m)\big) + \frac{4}{\kappa r_{m-1}} \cdot 17 r_m + 2C_3\varepsilon^2 + \frac{\varepsilon^2 r_m}{r_m}\right] \\ &\leq C\big[2\varepsilon^2 + 68\varepsilon^2 + 2C_3\varepsilon^2 + \varepsilon^2\big] \\ &\leq C'\varepsilon^2. \end{aligned} \tag{7.42}$$

For $t \in [\tau - r_m,\ \tau]$, define the interval

$$J_m(t) \doteq \big[\bar{x} - 8r_m + (t - t_m),\ \bar{x} + 8r_m - (t - t_m)\big].$$

Recalling that all wave propagation speeds are < 1, the standard interaction estimate (1.9) now yields

$$\begin{aligned} V\big(u(t);\ J_m(t)\big) &\leq V\big(u(t_m);\ J(t_m)\big) + C_1 V^2\big(u(t_m);\ J(t_m)\big) \\ &\leq C'\varepsilon^2 + C_1\big[C'\varepsilon^2\big]^2. \end{aligned}$$

This establishes (2.24) with $r^* \doteq r_N$. Indeed, this choice of r^* implies

$$[\bar{x} - 7r^* - r,\ \bar{x} + 7r^* + r] \subseteq J_N(\tau - r) \subseteq J_m(\tau - r) \quad m = 1,\ldots,N,\ r \in]0, r^*].$$

Now consider CASE 2. Let the intervals J, J' be as in (7.36). From the assumptions it now follows that, for some $r_0 > 0$, the function $\tilde{u}(t)$ has a shock of size $\tilde{\sigma}(t) < -\varepsilon^3/2$ at some point $\tilde{y}(t) \in J(t)$ for all $t \in [\tau - r_0,\ \tau]$. Moreover,

$$V\big(\tilde{u}(t);\ J'(t) \setminus \{y(t)\}\big) < \varepsilon^8/2 \qquad t \in [\tau - r_0,\ \tau].$$

Choose an integer N so large that $N\varepsilon^{16} > C_0$ and define

$$r_m \doteq (\varepsilon^{16}\kappa)^m r_0 \qquad t_m \doteq \tau - r_m, \qquad m = 0,\ldots,N. \tag{7.43}$$

Now consider a solution u suitably close to $\tilde{u}$. For some m we must have

$$Q\big(u(t_{m-1})\big) - Q\big(u(t_m)\big) < \varepsilon^{16}. \tag{7.44}$$

Observe that, if the distance

$$\sup_{t \in [\tau - r_0,\ \tau]} \big\|u(t) - \tilde{u}(t)\big\|_{\mathbf{L}^1} \tag{7.45}$$

is sufficiently small, then $u(t_m)$ will also have a shock of size $\sigma(t_m)$ at some point $y(t_m) \in J(t_m)$, with

$$\big|\sigma(t_m) - \tilde{\sigma}(t_m)\big| < \varepsilon^7, \qquad \big|y(t_m) - \tilde{y}(t_m)\big| < \varepsilon^7 \qquad t \in [t_{m-1}, t_m]. \tag{7.46}$$

Indeed, call $\tilde{u}^-$, $\tilde{u}^+$ respectively the left and right limits of $\tilde{u}(t_{m-1})$ around the shock. For every $\varepsilon_0 > 0$, if the quantity (7.45) is suitably small, there exists points $x_1 < \tilde{y}(t_{m-1}) < x_2$ such that

$$x_2 - x_1 < \varepsilon_0, \qquad \big|u(t_{m-1}, x_1) - \tilde{u}^-\big| < \varepsilon, \qquad \big|u(t_{m-1}, x_2) - \tilde{u}^+\big| < \varepsilon. \tag{7.47}$$

Hence, on the small interval $[x_1, x_2]$, the function $u(t_{m-1})$ contains waves which connect a state very close to $\tilde{u}^-$ with a state very close to $\tilde{u}^+$. Since all these waves are located within an interval of length $< \varepsilon_0$, if the interaction potential of these waves did not satisfy

$$Q\big(u(t_{m-1});\ [x_1, x_2]\big) < \varepsilon^{15}, \tag{7.48}$$

by choosing $\varepsilon_0 > 0$ suitably small a substantial amount of interaction would take place within the time interval $[t_{m-1}, t_m]$, in contradiction with (7.44). Hence (7.48) holds, and on the interval $[x_1, x_2]$ the function $u(t_{m-1})$ contains a shock satisfying (7.46), plus possibly other waves of small total strength. The remainder of the proof is similar to CASE 1.

In CASE 3, for some $r_0 > 0$ and all $t \in [\tau - r_0,\ \tau[$, on the interval $J(t)$ the function $\tilde{u}(t)$ will contain two shocks, say of sizes $\tilde{\sigma}_1(t)$, $\tilde{\sigma}_2(t)$. In this case, we first show that for any $\varepsilon_0 > 0$, every solution $u(t)$ suitably close to $\tilde{u}$ also contains two shocks of sizes σ_1, σ_2, with

$$\big|\sigma_i(t) - \tilde{\sigma}_i(t)\big| < \varepsilon_0 \qquad i = 1, 2,\ t \in [\tau - r^*,\ \tau].$$

Then we proceed as in CASE 1.

CHAPTER 8

Proof of Proposition 2.10

Proposition 2.10 is a consequence of the Lemmas 2.8, 2.11, 2.13 stated in Section 2. This entire section is thus devoted to proving these lemmas.

PROOF OF LEMMA 2.8. By assumption, every point (t, x) is contained in the interior of a stabilizing block. The conclusion will thus follow from the compactness of the set $[-M,\ M] \times [t^*, T]$. Indeed, let $\tilde{u}$ be a structurally stable ε-solution. Fix any $\tau \in [t^*, T]$. For each $x \in [-M, M]$ the conclusions of Proposition 2.7 hold at the point (τ, x), for some $r^* = r^*(x) > 0$. Choose finitely many points $x_1, \dots, x_\nu$ so that

$$[-M, M] \subseteq \bigcup_{j=1}^{\nu} \left[x_j - r^*(x_j),\ x_j + r^*(x_j)\right].$$

Define

$$\rho(\tau) \doteq \min_{j=1,\dots,\nu} r^*(x_j).$$

Repeat the same construction for each τ, then choose times $\tau_1 < \dots < \tau_N$ such that

$$[t^*, T] \subseteq \bigcup_{i=1}^{N} \left]\tau_i - \rho(\tau_i),\ \tau_i + \rho(\tau_i)\right[.$$

By inserting points $0 < t_0 < \tau_1 < t_1 < \dots < \tau_N < t_N = T$ with

$$\left[t_{i-1},\ t_i\right] \subseteq \left[\tau_i - \rho(\tau_i),\ \tau_i + \rho(\tau_i)\right],$$

all conclusions of Lemma 2.8 are satisfied.

PROOF OF LEMMA 2.13. It suffices to consider a regular path $\gamma\colon [a, b] \mapsto BV$, with each $u^\theta = \gamma(\theta)$ having the same number of jumps, say at $x_1^\theta < \dots < x_N^\theta$, with $x_\alpha^\theta \in [-M, M]$ for all θ, α. Moreover, let all functions u^θ coincide outside the interval $[-M, M]$. For a given $\nu \geq 1$, define

$$\theta_m \doteq a + \frac{m}{\nu}(b - a) \qquad m = 0, \dots, \nu.$$

If ν is sufficiently large, for each m there exist points p_α $(= p_{m,\alpha})$ such that

$$-M < p_0 < x_1^\theta < p_1 < \dots < p_{N-1} < x_N^\theta < p_N < M \qquad \forall \theta \in [\theta_{m-1}, \theta_m].$$

We now replace the restriction of the original path γ to $[\theta_{m-1}, \theta_m]$ with a new path γ' defined as follows. If $\vartheta \in [-M, p_0] \cup [p_N, M]$ we set

$$\gamma''(\vartheta) \doteq u^{\theta_m} \cdot \chi_{]-\infty,\ \vartheta]} + u^{\theta_{m-1}} \cdot \chi_{]\vartheta,\ \infty[}\,. \tag{8.1}$$

The same definition (8.1) is valid if $\vartheta \in]p_{\alpha-1}, p_\alpha]$ and $x_\alpha^{\theta_{m-1}} \le x_\alpha^{\theta_m}$. On the other hand, if $\vartheta \in]p_{\alpha-1}, p_\alpha]$ but $x_\alpha^{\theta_{m-1}} > x_\alpha^{\theta_m}$, we set

$$
\begin{aligned}
\gamma''(\vartheta) \doteq & u^{\theta_m} \cdot \chi_{]-\infty,\ p_{\alpha-1}]\ \cup\ ]p_{\alpha-1}+p_\alpha-\vartheta,\ p_\alpha]} \\
& + u^{\theta_{m-1}} \cdot \chi_{]p_{\alpha-1},\ p_{\alpha-1}+p_\alpha-\vartheta]\ \cup\ ]p_\alpha,\ \infty[}\,.
\end{aligned}
\tag{8.2}
$$

Clearly, γ'' is a piecewise regular path, with $\gamma''(-M) = \gamma(\theta_{m-1})$, $\gamma''(M) = \gamma(\theta_m)$. We now perform a suitable parameter rescaling: $\theta \mapsto \vartheta(\theta)$, mapping $[\theta_{m-1}, \theta_m]$ onto $[-M, M]$, and define the path

$$
\gamma'(\theta) \doteq \gamma''(\vartheta(\theta)) \qquad\qquad \theta \in [\theta_{m-1},\ \theta_m].
$$

Applying the same procedure to each subinterval $[\theta_{m-1}, \theta_m]$ we thus obtain a path $\gamma' \colon [a, b] \mapsto BV$ which has localized variation and coincides with γ at each point θ_m, $m = 0, \ldots, \nu$.

We can now consider a sequence of paths $(\gamma'_\nu)_{\nu \ge 1}$, constructed as above. Letting $\nu \to \infty$, from the regularity of the original path γ it follows

$$
\lim_{\nu\to\infty} \sup_{\theta\in[a,b]} \left\| \gamma'_\nu(\theta) - \gamma(\theta) \right\|_{\mathbf{L}^1} = 0, \qquad\qquad \lim_{\nu\to\infty} \|\gamma'_\nu\|_\star = \|\gamma\|_\star\,.
$$

Hence, choosing ν suitably large, all conclusions of Lemma 2.13 are satisfied.

PROOF OF LEMMA 2.11. As a preliminary, we give a formula for the weighted length of a tangent vector, providing the appropriate extension of (2.31)–(2.34) to the case of a function u with arbitrary jumps, not necessarily consisting of a single shock.

Let u be a piecewise Lipschitz function having jumps at the points $x_1 < \cdots < x_N$. Assume that the ε-solution of the Riemann problem determined by the jump at x_α consists of waves of sizes $\sigma_\alpha^1, \ldots, \sigma_\alpha^n$. Let the components of v, u_x be as in (2.30). The weighted norm of a generalized tangent vector $(v, \xi) \in T_u \doteq \mathbf{L}^1 \times \mathbb{R}^N$ is then defined as

$$
\left\|(v,\xi)\right\|_u^\star \doteq \sum_{i=1}^n \int_{-\infty}^{\infty} |v_i(x)| W_i^u(x)dx + \sum_{\alpha=1}^N \sum_{i=1}^n |\xi_\alpha| |\sigma_\alpha^i| W_i^u(x_\alpha). \tag{8.3}
$$

The weight $W_i^u(x)$ assigned to an i-wave located at x has the form

$$
W_i^u(x) \doteq 1 + \kappa_1 R_i^u(x) + \kappa_1\kappa_2 Q(u), \tag{8.4}
$$

where

$$
\begin{aligned}
R_i^u(x) \doteq & \left[\sum_{j\le i} \int_x^\infty + \sum_{j\ge i} \int_{-\infty}^x \right] |u_x^j(y)|dy + \left[\sum_{\substack{j\le i \\ x_\alpha > x}} + \sum_{\substack{j\ge i \\ x_\alpha < x}} \right] |\sigma_\alpha^j| \\
& + \begin{cases} \sigma_\alpha^i & \text{if } x = x_\alpha \text{ and } \sigma_\alpha^i > 0, \\ 0 & \text{otherwise.} \end{cases}
\end{aligned}
\tag{8.5}
$$

$$
\begin{aligned}
Q(u) \doteq & \sum_{i\le j} \iint_{x<y} |u_x^j(x)| |u_x^i(y)|\ dxdy\ +\ \sum_{\substack{i\le j \\ x_\alpha < x_\beta}} |\sigma_\alpha^j \sigma_\beta^i| + \sum_{\substack{i,\alpha \\ \sigma_\alpha^i > 0}} (\sigma_\alpha^i)^2 \\
& + \sum_{\alpha,j} \left[\sum_{i\le j} |\sigma_\alpha^j| \int_{x_\alpha}^\infty |u_x^i(x)|dx + \sum_{i\ge j} |\sigma_\alpha^j| \int_{-\infty}^{x_\alpha} |u_x^i(x)|dx \right],
\end{aligned}
\tag{8.6}
$$

and κ_1, κ_2 are suitably large constants. In terms of the measures μ_i introduced in Section 1, the quantities (8.5), (8.6) can be written as

$$R_i^u(x) \doteq \sum_{j\leq i} \mu_j(]x,\infty[) + \sum_{j\geq i} \mu_j(]-\infty, x[) + \big[\mu_i(\{x\})\big]_+, \tag{8.7}$$

$$Q(u) \doteq \sum_{i\leq j} \big(|\mu_j| \times |\mu_i|\big)\Big(\{(x,y);\ x<y\}\Big) + \sum_{\substack{i,x\\ \mu_i(\{x\})>0}} \mu_i(\{x\})^2. \tag{8.8}$$

The weighted length $\|\gamma\|_\star$ of a piecewise regular path and the weighted distance $d^\star(u,u')$ between two functions are then defined by (1.47), (1.48), respectively.

To simplify the notations, we assume that

$$\Gamma = \big\{(t,x);\ t\in[0,T],\ |x|\leq 4\rho - t\big\} \tag{8.9}$$

with $\rho > T$. Moreover, we assume that all functions $\bar u^\theta \doteq u^\theta(0,\cdot)$ coincide outside the interval $[-\rho,\ \rho]$ and satisfy one of the conclusions (i) or (ii) or (iii) in Proposition 2.7 on the interval $[-4\rho,\ 4\rho]$. In particular, the values

$$u^- \doteq \bar u^\theta(-4\rho) \qquad\qquad u^+ \doteq \bar u^\theta(4\rho) \tag{8.10}$$

are independent of θ. For each $\theta\in\Theta$, define the truncated functions

$$\hat u^\theta(x) \doteq \begin{cases} \bar u^\theta(x) & \text{if } x\in[-4\rho,\ 4\rho],\\ u^- & \text{if } x<-4\rho,\\ u^+ & \text{if } x>4\rho. \end{cases} \tag{8.11}$$

By the assumptions in Lemma 2.11, all functions $\hat u^\theta$ lie within a domain $\mathcal{D}$ consisting of:

CASE 1. Functions having no large shocks, and total variation $< C\varepsilon^2$.

CASE 2. Functions having exactly one large shock, say in the k-th family, and total variation (outside this shock) $< C\varepsilon^8$.

CASE 3. (a) Functions having exactly two large shocks, and total variation $< C\varepsilon^{20}$ outside these two shocks, together with (b) functions generated by the interaction of the two shocks in a solution of type (a).

Observe that, in Cases 1 and 2, our present domain $\mathcal{D}$ is a special case of the domains $\mathcal{D}_{(u^-,u^+)}$ considered in Proposition 2.4. The constructive procedures developed in Sections 3, 4 can thus be used. Case 3 is somewhat different, and the construction of piecewise Lipschitz approximations therefore needs to be suitably modified, taking into account the possible interaction of the two large shocks. In the following, we will work out a detailed proof of Lemma 2.11 in Case 2. The same arguments can be easily adapted to Case 1, which is much easier. The modifications needed to cover Case 3 will be discussed at the end of the section.

In the main part of the proof, we show that the weighted length of the path $\gamma_t\colon \theta\mapsto S_t^\varepsilon u^\theta$ does not increase in time, under the additional assumption that $\bar u^\theta = \hat u^\theta$ for all θ. Afterwards, we show that the result remains true in the general case, relying on the fact that all functions u^θ coincide outside the region Γ.

The basic idea is to construct a path $\tilde\gamma\colon \theta\mapsto w^\theta$ of approximate ε-solutions such that:

- The weighted length of the path $\tilde{\gamma}_0 \colon \theta \mapsto w^\theta(0,\cdot)$ is arbitrarily close to the weighted length of the path $\gamma_0 \colon \theta \mapsto \bar{u}^\theta$.
- For all $t \in [0,T]$ and $\theta \in \Theta$, the distance $\left\|w^\theta(t,\cdot) - u^\theta(t,\cdot)\right\|_{\mathbf{L}^1}$ remains arbitrarily small.
- The increase in the weighted length of the path $\tilde{\gamma}_t \colon \theta \mapsto w^\theta(t,\cdot)$ is arbitrarily small, as t varies from 0 to T.

We remark that, for a path $\tilde{\gamma}$ of piecewise Lipschitz approximate solutions constructed as in Sections 3-4, the increase in the weighted length $\|\tilde{\gamma}_t\|_\star$ defined by (3.42)–(3.45) can be kept arbitrarily small. Our present goal is to show that the same result holds if (3.32) and (3.44) are replaced respectively by (8.6) and (8.5), provided that the width δ_2 of the shock layers in (3.5) is sufficiently small. The proof will be accomplished by first reducing the problem to a few special cases.

1. Observe that a family of approximate ε-solutions of (1.1) can be obtained by choosing a time step $\Delta t = \delta_1 > 0$ and performing a cyclical concatenation of n distinct flows:

$$w^\theta(t) = \cdots S^2_{\Delta t} \circ S^1_{\Delta t} \circ S^n_{\Delta t} \circ \cdots \circ S^2_{\Delta t} \circ S^1_{\Delta t} \bar{w}^\theta, \tag{8.12}$$

where each S^h is a semigroup related to the hyperbolic system (1.44), with $n-1$ linearly degenerate fields. For $h = 1,\ldots,n$, a convenient way to define the semigroup $S^h \colon \mathcal{D} \times [0,\infty[\mapsto \mathcal{D}$ is to specify how it acts on piecewise Lipschitz data. This is done as follows.

Let $\mathcal{D}$ be as in (2.8), with $\eta = O(\varepsilon^8)$. Observe that every $u \in \mathcal{D}$ has a single large k-shock, say located at the point y_k. Introduce the speeds

$$\begin{aligned} \lambda_i^* &= \lambda_i(u^-) \qquad\qquad i \neq k, \\ \lambda_k^* &= \lambda_k(u^-),\ \ \lambda_k^{**} = \lambda_k(u^+). \end{aligned} \tag{8.13}$$

We are of course in a special case of (3.1), with $\mathcal{S} = \{k\}$. Consider the hyperbolic system

$$u_t + A_h(x,u)u_x = 0 \tag{8.14}$$

where A_h is the matrix with the same eigenvectors as $A(u)$, and whose eigenvalues $\lambda_1^{(h)},\ldots,\lambda_n^{(h)}$ are as follows:

$$\lambda_i^{(h)} = \begin{cases} \lambda_i^* & \text{if } i \neq k, \\ \lambda_i^* & \text{if } i = k,\ x < y_k, \\ \lambda_i^{**} & \text{if } i = k,\ x > y_k, \end{cases} \tag{8.15}$$

in the case $i \neq h$, while

$$\lambda_h^{(h)} = \lambda_h^* + n\big(\lambda_h(u) - \lambda_h^*\big) \tag{8.16}$$

if $h \neq k$, and

$$\lambda_h^{(h)} = \begin{cases} \lambda_h^* + n\big(\lambda_h(u) - \lambda_h^*\big) & \text{if } x < y_k, \\ \lambda_h^{**} + n\big(\lambda_h(u) - \lambda_h^{**}\big) & \text{if } x > y_k \end{cases} \tag{8.17}$$

if $h = k$.

Concerning the shocks, we require that the single large jump at y_k satisfies the ε-Rankine Hugoniot equations (2.3)–(2.5) with $i = k$. Every small shock, say of the

j-th family, located at y_α, satisfies the relations

$$u(y_\alpha+) = R_j(\sigma_\alpha)\big(u(y_\alpha-)\big), \tag{8.18}$$

for some wave size σ_α. If $j \neq h$, the speed of a small j-shock is

$$\dot{y}_\alpha = \begin{cases} \lambda_j^* & \text{if } j \neq k, \\ \lambda_j^* & \text{if } j = k,\ y_\alpha < y_k, \\ \lambda_j^{**} & \text{if } j = k,\ y_\alpha > y_k. \end{cases} \tag{8.19}$$

Finally, the speed of a small h-shock is

$$\dot{y}_\alpha = (1-n)\lambda_h^* + \frac{n}{|\sigma_\alpha|}\int_{\sigma_\alpha}^0 \lambda_h\Big(R_h(s)\big(u(y_\alpha-)\big)\Big)ds, \tag{8.20}$$

if $h \neq k$ or if $h = k$, $y_\alpha < y_k$, while

$$\dot{y}_\alpha = (1-n)\lambda_h^{**} + \frac{n}{|\sigma_\alpha|}\int_{\sigma_\alpha}^0 \lambda_h\Big(R_h(s)\big(u(y_\alpha-)\big)\Big)ds, \tag{8.21}$$

if $h = k$, $y_\alpha > y_k$. We now define S^h as the unique Lipschitz semigroup with domain $\mathcal{D}$ as in (2.8) with the following property. If $\bar{u} \in \mathcal{D}$ is piecewise constant, then for $t > 0$ small $S_t^h\bar{u}$ is the unique piecewise Lipschitz function which satisfies the quasilinear hyperbolic system (8.14) a.e., together with the relations (8.18)–(8.21) along the shock lines.

Observe that, if in the approximations constructed in Sections 3-4 we vary the width of the shock layer $\delta_2 > 0$ and keep the time step $\delta_1 > 0$ fixed, then on each subinterval $I_{m,h}$ $(m \geq 0)$ at (3.2) in the limit $\delta_2 \to 0$ we obtain precisely the flow of S^h. The existence and uniqueness of the semigroup S^h thus follows from the analysis in Sections 3–5 as a special case. If we can show that each S^h is contractive for the weighted distance determined by (8.3)–(8.6), the same will of course hold for every concatenation of the form (8.12). Letting $\Delta t \to 0$, by possibly taking a convergent subsequence, we thus obtain a Lipschitz semigroup $\widehat{S}$. The same arguments used in the proof of Proposition 2.4 now show that $\widehat{S}_t\bar{u} = S_t^\varepsilon\bar{u}$ for every piecewise constant $\bar{u}$ and every $t > 0$ small enough. By uniqueness, $\widehat{S} = S^\varepsilon$. In particular, by taking $\Delta t > 0$ small enough, we can assume that the distance between the flow of S^ε and the corresponding concatenation of flows in (8.12) is as small as we like.

The proof is thus reduced to showing that the flow of each semigroup S^h, $h = 1, \ldots, n$, is contractive w.r.t. the metric (8.3)–(8.6). More precisely, given a regular path of initial data $\gamma_0 \colon \theta \mapsto \bar{u}^\theta \in \mathcal{D}$, for each $\varepsilon' > 0$ it suffices to exhibit a path $\tilde{\gamma}_T \colon \theta \mapsto w^\theta(T, \cdot)$ such that

$$\|\tilde{\gamma}_T\|_\star \leq \|\gamma_0\|_\star + \varepsilon', \tag{8.22}$$

$$\left\|\tilde{\gamma}_T(\theta) - S_T^h\bar{u}^\theta\right\|_{\mathbf{L}^1} < \varepsilon' \qquad \theta \in \Theta. \tag{8.23}$$

2. As a second reduction, we observe that any path γ can be approximated by a path γ' having localized variation. More precisely, γ' can be chosen to be a finite concatenation of paths of the form (8.1) or (8.2). Furthermore, any path of the form (8.1) or (8.2) can be approximated by another path where u^{θ_m}, $u^{\theta_{m-1}}$ are piecewise constant.

Now consider a path of the form

$$\gamma(\theta) = u \cdot \chi_{]-\infty,\theta]} + u' \cdot \chi_{]\theta,\infty[} \qquad \theta \in \Theta \doteq [a,b], \tag{8.24}$$

assuming that u, u' are piecewise constant and $u(x) = u^\flat$, $u'(x) = u^\sharp$ for all $x \in [a,b]$. Let $\omega_0 = u^\flat$, ω_1, ... , $\omega_n = u^\sharp$ be the states determined by the ε-solution of the Riemann problem with data $u^\flat, u^\sharp$. Then the concatenation $\gamma_n \circ \cdots \circ \gamma_1$, with

$$\gamma_i(\theta) = u \cdot \chi_{]-\infty,a]} + \omega_{i-1} \cdot \chi_{]a,\theta]} + \omega_i \cdot \chi_{]\theta,b]} + u' \cdot \chi_{]b,\infty]}$$

has the same weighted length as the original path γ.

3. Thanks to the above remarks, it suffices to construct a path of approximate solutions $\tilde{\gamma}\colon \theta \mapsto w^\theta$, satisfying (8.22)-(8.23), under the following additional assumptions:

(A) The path $\gamma_0\colon \theta \mapsto \bar{u}^\theta$ has the form (8.24), with u, u' piecewise constant, $u(x) = u^\flat$, $u'(x) = u^\sharp$ for $x \in [a,b]$, and each Riemann problem determined by a jump in $\bar{u}^\theta$ is solved by a single wave. Moreover, $[a,b] \subseteq [-\rho,\ \rho]$.

We shall distinguish four cases, assuming that the jump $(u^\flat, u^\sharp)$ is

(i) the large k-shock,
(ii) a small h-shock,
(iii) a small j-rarefaction, for some $j \in \{1, \ldots, n\}$,
(iv) a small j-shock, for some $j \neq h$.

In all cases, we insert a thin shock layer of width $\delta_2 > 0$ around the large k-shock, say

$$[y_k^*,\ y_k^{**}] \doteq [y_k - \delta_2,\ y_k + \delta_2], \tag{8.25}$$

and construct piecewise Lipschitz approximate solutions following the same procedure used in Sections 3-4. Observe that we are here in the special case where $\mathcal{S} = \{k\}$ and all the small shocks belong to the h-th family. A straightforward application of the estimates (3.50)–(3.71) on tangent vectors, however, is not possible. Indeed, the weighted norm introduced at (8.3)–(8.6) is different from the norm (3.42)–(3.45) because:

- waves of the same family are now always regarded as approaching,
- in (8.5) there is no term related to the shock layer around the big shock.

To keep track of how the weighted length of a path $\gamma_t\colon \theta \mapsto u^\theta(t,\cdot)$ changes in time, we consider an auxiliary weighted norm, defined as follows. Let u be piecewise Lipschitz with jumps at points y_α, $\alpha \in \mathcal{S} \cup \mathcal{S}' = \{k\} \cup \mathcal{S}'$. To fix the ideas, let $h > k$, the other cases being similar. Let u have a small h-shock of strength $|\sigma_\alpha|$ at each point y_α, $\alpha \in \mathcal{S}'$, together with a large k-shock at y_k. Let y_k^*, y_k^{**} be as in (8.25). We then define

$$\begin{aligned} \big\|(v,\xi)\big\|_u^\diamond \doteq & \sum_{i=1}^n \int_{-\infty}^{\infty} \widetilde{W}_i^u(x)\big|v_i(x)\big|dx + \sum_{\alpha\in\mathcal{S}\cup\mathcal{S}'} \widetilde{W}_{k_\alpha}^u |\sigma_\alpha||\xi_\alpha| \\ & - \varepsilon^3 \int_{y_k^*}^{y_k} \big|v_h(x)\big|\ dx, \end{aligned} \tag{8.26}$$

$$\widetilde{W}_i^u(x) \doteq 1 + \kappa_1 \widetilde{R}_i^u(x) + \kappa_1\kappa_2 Q(u). \tag{8.27}$$

Here $Q(u)$ is the interaction potential (8.6), while for $x \neq y_k$ we set

$$
\widetilde{R}_i^u(x) \doteq \left[\sum_{j\le i}\int_x^{\infty} + \sum_{j\ge i}\int_{-\infty}^{x}\right] \left|u_x^j(y)\right| dy + \left[\sum_{\substack{\alpha\in S\cup S' \\ k_\alpha\le i,\ y_\alpha>x}} + \sum_{\substack{\alpha\in S\cup S' \\ k_\alpha\ge i,\ y_\alpha<x}}\right] |\sigma_\alpha|
$$

$$
+\begin{cases} 0 & \text{if } x\notin [y_k^*,\ y_k^{**}], \\ -\varepsilon^3 & \text{if } \quad x\in [y_k^*,\ y_k[\,,\ i\ge k, \text{ or if } x\in\,]y_k,\ y_k^{**}],\ i\le k, \\ \varepsilon^3 & \text{if } x\in [y_k^*,\ y_k[\,,\ i<k, \text{ or if } x\in\,]y_k,\ y_k^{**}],\ i>k. \end{cases} \tag{8.28}
$$

For the large k-shock located at y_k we define

$$
\begin{aligned}
\widetilde{R}_k^u(y_k) \doteq{}& \left[\sum_{j\le k}\int_{y_k}^{\infty} + \sum_{j\ge k}\int_{-\infty}^{y_k}\right] \left|u_x^j(y)\right| dy \\
&+ \left[\sum_{\substack{\alpha\in S' \\ k_\alpha\le k,\ y_\alpha>y_k}} + \sum_{\substack{\alpha\in S' \\ k_\alpha\ge k,\ y_\alpha<y_k}}\right] |\sigma_\alpha| \\
&+\varepsilon\cdot\left\{\int_{y_k^*}^{y_k}\Big(\sum_{j<k}\left|u_x^j(x)\right| - \sum_{j\ge k}\left|u_x^j(x)\right|\Big)dx\right. \\
&\qquad + \int_{y_k}^{y_k^{**}}\Big(\sum_{j>k}\left|u_x^j(x)\right| - \sum_{j\le k}\left|u_x^j(x)\right|\Big)dx \\
&\qquad \left. + \sum_{\substack{\beta\in S' \\ y_\beta\in\,]y_k,\,y_k^{**}]}} |\sigma_\beta| - \sum_{\substack{\beta\in S' \\ y_\beta\in[y_k^*,\,y_k[}} |\sigma_\beta|\right\}.
\end{aligned} \tag{8.29}
$$

The weighted length of a path γ corresponding to the metric (8.26)–(8.29) will be denoted by $\|\gamma\|_\diamond$, while $\|\gamma\|_\star$ always refers to the metric (8.3)–(8.6). The definitions (8.26)–(8.29) are chosen so that, for a fixed u and a tangent vector (v,ξ), letting $\delta_2\to 0$ we have $y_k^*, y_k^{**}\to y_k$ and hence

$$
\lim_{\delta_2\to 0}\left\|(v,\xi)\right\|_u^\diamond = \left\|(v,\xi)\right\|_u^\star .
$$

4. Let the assumptions (A) hold. We then construct approximate solutions w^θ as in Sections 3-4. During a time interval between two consecutive restarting times, a minor modification of the estimates (3.50)–(3.71) shows that the weighted length $\|\tilde\gamma_t\|_\diamond$ of the path $\tilde\gamma_t\colon \theta\mapsto w^\theta(t,\cdot)$ is a non-increasing function of time. Let us examine in more detail what happens at a time t where a restarting algorithm is used.

- When a steep compressive h-wave is replaced by several small h-shocks, the weighted norm of tangent vectors decreases. Indeed, when a family of negative h-waves are collapsed into a single point, they are no longer regarded as approaching each other.

- When a small h-shock penetrates inside the shock layer around the big shock at y_k, it is replaced by a steep compression wave. In this case, the weighted norm of tangent vectors still decreases, because of the last term in (8.26).

- At the initial time $t = 0$, we need to replace each piecewise constant function $u^\theta(0,\cdot) = \bar u^\theta$ by some other function $w^\theta(0,\cdot)$ having one large k-shock, possibly several small h-shocks, and no other jumps. More precisely, let u^θ have a j_α-jump of size σ_α at each point y_α (independent of θ), a large k-shock at the point y_k, and a j-shock of strength $|\bar\sigma|$ at $x = \theta$. The restarting procedure will replace each jump with $j_\alpha \neq h$ with a continuous wave. In a neighborhood of each point y_α, if $j_\alpha \neq h$, the new function $\tilde\gamma_0(\theta) = w^\theta(0,\cdot)$ will have the form

$$w^\theta(0,x) = \begin{cases} \bar u^\theta(y_\alpha -) & \text{if } x < y_\alpha, \\ R_{j_\alpha}\big((x-y_\alpha)\sigma_\alpha/\delta^*\big)\big(\bar u^\theta(y_\alpha -)\big) & \text{if } x \in [y_\alpha,\ y_\alpha+\delta^*], \\ \bar u(y_\alpha +) & \text{if } x > y_\alpha + \delta^*. \end{cases} \tag{8.30}$$

Here R_j denotes a j-rarefaction curve and $\delta^* > 0$ is a suitably small constant. Similarly, in a neighborhood of the point $x = \theta$, in cases (iii) and (iv) the function $w^\theta(0,\cdot)$ is given by

$$w^\theta(0,x) = \begin{cases} u^\flat & \text{if } x < \theta, \\ R_j\big((x-\theta)\bar\sigma/\delta^*\big)(u^\flat) & \text{if } x \in [\theta,\ \theta+\delta^*], \\ u^\sharp & \text{if } x > \theta+\delta^*. \end{cases} \tag{8.31}$$

Here $\bar\sigma$ is the size of the jump $(u^\flat, u^\sharp)$, at $x = \theta$. If $j_\alpha = h$, then on a neighborhood of y_α we simply have $w^\theta(0,x) = \bar u^\theta(x)$. The same holds on a neighborhood of the big k-shock at y_k. Since u^θ does not change with θ outside $[a,b]$, by (8.31), in cases (iii)-(iv) the generalized tangent vector (v,ξ) to the path $\tilde\gamma_0$ is computed by

$$v_j = -\frac{\bar\sigma}{\delta^*}\cdot \chi_{[\theta,\ \theta+\delta^*]}, \qquad v_i \equiv 0 \text{ for } i \neq j, \qquad \xi \equiv 0. \tag{8.32}$$

In cases (i), (ii) and (iii), choosing δ_2 sufficiently small, the weighted length of the new path $\tilde\gamma_0 \colon \theta \mapsto w^\theta(0,\cdot)$ is

$$\|\tilde\gamma_0\|_\diamond = \|\tilde\gamma_0\|_\star = \|\gamma_0\|_\star + \kappa_1\kappa_2(b-a)|\bar\sigma|\cdot \sum_{\substack{j_\alpha\neq h,\\ \sigma_\alpha<0}} |\sigma_\alpha|^2, \tag{8.33}$$

where $|\bar\sigma|$ is the strength of the jump at $x = \theta$. The increase is due to the change in Q at (8.6). In case (iv) we have

$$\begin{aligned} \|\tilde\gamma_0\|_\diamond &= \|\tilde\gamma_0\|_\star \\ &= \|\gamma_0\|_\star + \kappa_1(b-a)|\bar\sigma|^2 + \kappa_1\kappa_2(b-a)|\bar\sigma|\left(|\bar\sigma|^2 + \sum_{\substack{j_\alpha\neq h,\\ \sigma_\alpha<0}} |\sigma_\alpha|^2\right). \end{aligned} \tag{8.34}$$

For any given $\varepsilon' > 0$, if δ_2, δ^* in (8.25) and (8.30)-(8.31) are small enough and if all restarting procedures are suitably accurate, we then have

$$\left\|\tilde\gamma_t(\theta) - S^h_T\bar u^\theta\right\|_{\mathbf{L}^1} < \varepsilon' \qquad \text{for all } \theta\in[a,b],\ |t-T|<\delta, \tag{8.35}$$

for some $\delta > 0$. Our main concern is thus to control the weighted length of the paths $\tilde\gamma_t \colon \theta \mapsto w^\theta(t,\cdot)$.

5. Before proceeding with the proof, let us point out the two main difficulties that we are facing.

(a) By (8.33) or (8.34), in general one has $\|\tilde{\gamma}_0\|_\diamond = \|\tilde{\gamma}_0\|_\star > \|\gamma_0\|_\star$. This is because the restarting performed at the initial time $t = 0$ "spreads out" all small j-shocks $(j \neq h)$, thus increasing the length of the path.
To compensate for this, at the terminal time $t = T$ we shall replace the steep compressive j-waves $(j \neq h)$ with j-shocks. This will shorten the weighted length of the path $\tilde{\gamma}_T$ by the appropriate amount.

(b) The norms $\|(v,\xi)\|_w^\diamond$ and $\|(v,\xi)\|_w^\star$ are constructed assigning different weights to waves inside the shock layer $[y_k^*, y_k^{**}]$ in (8.25). Hence, the weighted lengths $\|\tilde{\gamma}_T\|_\diamond$ and $\|\tilde{\gamma}_T\|_\star$ may be substantially different.
Roughly speaking, this difficulty is overcome by choosing a time $\tau \in [T,\ T+\varepsilon_0]$ when "most part" of the tangent vector (v^θ, ξ^θ) has already flowed out from the shock layer $[y_k^\theta - \delta_2,\ y_k^\theta + \delta_2]$, for most values of θ. In this way, the weighted lengths $\|\tilde{\gamma}_\tau\|_\diamond$ and $\|\tilde{\gamma}_\tau\|_\star$ will be almost the same.

We first address **(b)**. Observe that, for any given $\varepsilon' > 0$ and $\delta > 0$ as in (8.35), we can choose $\varepsilon_0 \in]0, \delta]$ with the following property. If

$$\sum_{i=1}^{n} \int_{y_k^*}^{y_k^{**}} \left|w_x^i(x)\right| dx + \sum_{y_\alpha \in [y_k^*, y_k^{**}]} |\sigma_\alpha| < \varepsilon_0, \tag{8.36}$$

$$\sum_{i=1}^{n} \int_{y_k^*}^{y_k^{**}} \left|v_i(x)\right| dx + \sum_{y_\alpha \in [y_k^*, y_k^{**}]} |\sigma_\alpha||\xi_\alpha| < \varepsilon_0 \tag{8.37}$$

for all $\theta \notin \Theta'$, for some set $\Theta' \subset \Theta$ with

$$\text{meas}(\Theta') < \varepsilon_0, \tag{8.38}$$

then

$$\|\gamma\|_\star < \|\gamma\|_\diamond + \varepsilon'. \tag{8.39}$$

In (8.36)-(8.37), it is understood that $w = w^\theta(t,\cdot) = \gamma_t(\theta)$, while $(v,\xi) = (v^\theta, \xi^\theta)$ is the corresponding tangent vector. To establish (8.22), we thus need to prove the following

CLAIM 8.1. *For any given $\varepsilon_0 > 0$, by choosing $\delta_2 > 0$ sufficiently small, there exists at least one time $\tau \in [T,\ T+\varepsilon_0]$ such that the corresponding estimates (8.36)-(8.37) hold for all θ in a set $\Theta \subset [a,b]$ satisfying (8.38).*

Toward a proof of the above claim, the key observation is that all waves of all characteristic families cross the boundary region

$$B \doteq [y_k - \delta_2,\ y_k[\ \cup\ ]y_k,\ y_k + \delta_2]$$

transversally. Roughly speaking, if some function $w^\theta(T,\cdot)$ contains a large amount of waves inside the shock layer B, we can simply wait until a later time $T + \Delta t$, with $\Delta t = O(\delta_2)$, when all these waves will have moved outside B. Since ε_0 is fixed and δ_2 can be chosen arbitrarily small, our claim holds. We now turn this intuitive argument into a rigorous proof.

Recalling (1.9), call

$$\Upsilon_\theta(t) \doteq V\big(w^\theta(t)\big) + C_1 \cdot Q\big(w^\theta(t)\big),$$

so that the quantity $\Upsilon_\theta(t_1) - \Upsilon_\theta(t_2)$ provides an upper bound for the total amount of interaction and cancellation in the solution w^θ during the time interval $[t_1, t_2]$. Moreover, define

$$\Upsilon'_\theta(t) \doteq \Big\| \big(v^\theta(t),\ \xi^\theta(t)\big) \Big\|^\diamond_{w^\theta(t)}.$$

Observe that, for every fixed θ, the positive variation of both functions $\Upsilon_\theta, \Upsilon'_\theta$ can be made arbitrarily small by increasing the accuracy of the restarting procedure.

In the following, we fix a lower bound $\Delta\lambda > 0$ for the absolute value of the difference between the speed of the large shock at y_k and every other wave speed, for every function w^θ. Given any solution w^θ, if

$$x \in \big[y_k(t) - \delta_2,\ y_k(t) + \delta_2\big],$$

then for every $i \in \{1, \ldots, n\}$ either the forward or the backward generalized i-characteristic through (t, x) crosses the big k-shock at some time

$$t' \in I_t \doteq [t - \delta_2/\Delta\lambda,\ t + \delta_2/\Delta\lambda].$$

In particular, if (8.36) does not hold for w^θ at time t, then a uniformly positive amount of interaction and cancellation must take place within the time interval I_t, hence

$$\Upsilon_\theta(t - \delta_2/\Delta\lambda) - \Upsilon_\theta(t + \delta_2/\Delta\lambda) > \kappa\varepsilon_0, \tag{8.40}$$

for some constant $\kappa > 0$. Similarly, if (8.37) fails, then

$$\Upsilon'_\theta(t - \delta_2/\Delta\lambda) - \Upsilon'_\theta(t + \delta_2/\Delta\lambda) > \kappa\varepsilon_0. \tag{8.41}$$

Let K be a constant such that

$$V(u) + C_1 \cdot Q(u) < K$$

for all functions u in the domain of the semigroup, and such that

$$\Upsilon'_\theta(0) = \Big\| \big(v^\theta(0), \xi^\theta(0)\big) \Big\|^\diamond_{w^\theta(0)} < K \qquad \forall \theta \in [a, b].$$

Choose a large integer N and a value $\delta_2 > 0$ so small that

$$N > \frac{2K(b-a)}{\kappa\varepsilon_0^2}, \qquad \frac{2\delta_2 N}{\Delta\lambda} < \varepsilon_0. \tag{8.42}$$

By the second inequality in (8.42), the interval $[T,\ T + \varepsilon_0]$ contains N disjoint subintervals of the form

$$[t_i^-,\ t_i^+] \doteq \big[t_i - \delta_2/\Delta\lambda,\ t_i + \delta_2/\Delta\lambda\big].$$

Assume that, for every t_i, (8.36) fails on a set Θ_i and (8.37) fails on a set Θ'_i, with

$$\text{meas}(\Theta_i) + \text{meas}(\Theta'_i) \geq \varepsilon_0.$$

In this case we would have

$$\Upsilon_\theta(t_i^-) - \Upsilon_\theta(t_i^+) > \kappa\varepsilon_0 \qquad \Theta \in \Theta_i,$$

$$\Upsilon'_\theta(t_i^-) - \Upsilon'_\theta(t_i^+) > \kappa\varepsilon_0 \qquad \Theta \in \Theta'_i.$$

Therefore,

$$\begin{aligned}2K(b-a) &\geq \int_a^b \Upsilon_\theta(T) - \Upsilon_\theta(T+\varepsilon_0)\,d\theta + \int_a^b \Upsilon'_\theta(T) - \Upsilon'_\theta(T+\varepsilon_0)\,d\theta \\ &\geq \sum_{i=1}^N \int_a^b \Upsilon_\theta(t_i^-) - \Upsilon_\theta(t_i^+)\,d\theta + \sum_{i=1}^N \int_a^b \Upsilon'_\theta(t_i^-) - \Upsilon'_\theta(t_i^+)\,d\theta \\ &\geq \sum_{i=1}^N \kappa\varepsilon_0 \cdot \big(\mathrm{meas}(\Theta_i) + \mathrm{meas}(\Theta'_i)\big) \\ &\geq N\kappa\varepsilon_0^2,\end{aligned}$$

in contradiction with the choice of N in (8.42). Hence, at some $t = t_i$, both (8.36) and (8.37) hold for θ in the set

$$\Theta \doteq [a,b] \setminus (\Theta_i \cup \Theta'_i)$$

satisfying (8.38). This proves the claim.

By the previous analysis, for every $\varepsilon' > 0$ we have shown the existence of a path $\tilde\gamma_\tau : \theta \mapsto w^\theta(\tau,\cdot)$ such that

$$\|\tilde\gamma_\tau\|_\star \leq \|\tilde\gamma_0\|_\star + \varepsilon', \qquad \left\|\tilde\gamma_\tau(\theta) - S_T^h \bar u^\theta\right\|_{\mathbf{L}^1} < \varepsilon' \qquad \theta \in \Theta. \tag{8.43}$$

6. If now $\|\tilde\gamma_0\|_\star = \|\gamma_0\|_\star$, i.e. if the initial functions $\bar u^\theta$ do not contain any small i-shock, for all $i \neq h$, then the proof is completed. In the general case, we need to show that, at some terminal time $\tau \in [T,\ T+\varepsilon_0]$, one can collapse the i-compression waves ($i \neq h$) back into a single shock and reduce the length of the path $\tilde\gamma_\tau$ by the appropriate amount.

To fix the ideas, let case (iii) hold, so that the jump $(u^\flat, u^\sharp)$ is solved by a single j-rarefaction, say with $j > k$, and (8.33) holds. The other cases can be handled by similar techniques. For $i \neq h$, consider the linear equation

$$z_t + \lambda_i^{(h)} z_x = 0. \tag{8.44}$$

Recall that, as in (3.4),

$$\lambda_i^{(h)} = \begin{cases} \lambda_i^* & \text{if } i \neq k, \\ \lambda_i^* & \text{if } i = h,\ x < y_k(t), \\ \lambda_i^{**} & \text{if } i = h,\ x > y_k(t). \end{cases}$$

We now introduce the auxiliary functions $\hat v_i, \hat w_x^i$ (also depending on θ), defined as follows. For $i \neq h$, $\hat v_i$ is the solution of (8.44) with initial data

$$\hat v_i(0,x) = \begin{cases} 0 & \text{if } i \neq j, \\ -(\bar\sigma/\delta^*) \cdot \chi_{[\theta,\ \theta+\delta^*]}(x) & \text{if } i = j, \end{cases} \tag{8.45}$$

while $\hat w_x^i$ is the solution of (8.44) with initial data

$$\hat w_x^i(0,x) = w_x^i(0,x). \tag{8.46}$$

Moreover, we set

$$\hat v_h \equiv 0, \qquad \hat w_x^h \equiv 0, \tag{8.47}$$

$$\tilde v_i \doteq v_i - \hat v_i, \qquad \tilde{w}_x^i \doteq w_x^i - \hat w_x^i \qquad i = 1, \ldots, n. \tag{8.48}$$

Consider the auxiliary weighted norm $\big\|(v,\xi)\big\|_w^\heartsuit$, defined as in (8.26)–(8.29) and (8.6), replacing the quantity $|v_i|$ by $|\tilde v_i| + |\hat v_i|$ and $|w_x^i|$ by $|\tilde w_x^i| + |\hat w_x^i|$ throughout. In particular

$$\big\|(v,\xi)\big\|_u^\heartsuit \doteq \sum_{i=1}^n \int_{-\infty}^{\infty} \widetilde{W}_i^w(x)\Big(\big|\tilde v_i(x)\big| + \big|\hat v_i(x)\big|\Big)dx$$

$$+ \sum_{\alpha\in\mathcal{S}\cup\mathcal{S}'} \widetilde{W}_{k_\alpha}^w |\sigma_\alpha|\big|\xi_\alpha\big| - \varepsilon^3 \int_{y_k^*}^{y_k} \big|v_h(x)\big|\, dx,$$

where the weights $\widetilde{W}_i^w$ are obtained from (8.28)-(8.29) and (8.6) with the due replacements.

Call $\|\tilde\gamma\|_\heartsuit$ the corresponding weighted length of the path $\tilde\gamma$. As in the previous cases, the positive variation of the function $t \mapsto \|\tilde\gamma_t\|_\heartsuit$ can be made arbitrarily small by increasing the accuracy of the restarting procedures. In particular, for any given $\varepsilon' > 0$ we can assume

$$\|\tilde\gamma_t\|_\heartsuit \le \|\tilde\gamma_0\|_\heartsuit + \varepsilon' \qquad t > 0. \tag{8.49}$$

At the initial time we have

$$\hat v_h = 0, \quad \hat w_x^h = 0, \qquad \tilde v_i = 0,\ \tilde w_x^i = 0 \quad \text{if} \quad i \neq h.$$

Therefore,

$$\|\tilde\gamma_0\|_\diamondsuit = \|\tilde\gamma_0\|_\heartsuit. \tag{8.50}$$

We now show that, at any time τ, from $\tilde\gamma_\tau$ one can construct a shorter path $\bar\gamma_\tau$ by collapsing each compressive i-wave in $w^\theta(\tau,\cdot)$ into a single point. Let $\{y_\alpha;\ \alpha\in\mathcal{S}''\}$ be the set of points (independent of θ) where the initial functions $\bar u^\theta$ have a small shock, in some family $j_\alpha \neq h$. By construction, $w^\theta(0,\cdot)$ will thus have a steep compressive j_α-wave on each interval $[y_\alpha,\ y_\alpha+\delta^*]$. At a given time $\tau > 0$, consider the intervals

$$J_\alpha \doteq [y_\alpha + \lambda_\alpha\tau,\ y_\alpha + \lambda_\alpha\tau + \delta^*], \tag{8.51}$$

where

$$\lambda_\alpha \doteq \begin{cases} \lambda_{j_\alpha}^* & \text{if } j_\alpha \neq k, \\ \lambda_k^* & \text{if } j_\alpha = k,\ y_\alpha < y_k, \\ \lambda_k^{**} & \text{if } j_\alpha = k,\ y_\alpha > y_k. \end{cases}$$

Moreover, call

$$J \doteq \bigcup_{\alpha\in\mathcal{S}''} J_\alpha. \tag{8.52}$$

The new path $\bar\gamma_\tau : \theta \mapsto \bar w^\theta(\tau,\cdot)$ is now obtained by collapsing each interval J_α into a single point. More precisely, we define $\bar w^\theta$ implicitly by setting

$$\bar w^\theta\Big(\tau,\ x - \text{meas}\big(J\setminus(-\infty,x]\big)\Big) \doteq w^\theta(\tau,x). \tag{8.53}$$

Observe that, for any given $\varepsilon_0 > 0$, by choosing δ^* sufficiently small there exists some time $\tau\in[T,\ T+\varepsilon_0]$ where all intervals J_α are disjoint. Moreover, we can assume that there exists a small set of parameters $\Theta' \subset \Theta$ satisfying (8.38), such that the relations (8.36)-(8.37) hold together with

$$y_k^\theta(\tau) \notin J \tag{8.54}$$

for all $\theta \notin \Theta'$. We now estimate the weighted length of the new path $\bar{\gamma}_\tau$. For $\theta \notin \Theta'$, define

$$\mathcal{S}''_k \doteq \Big\{\alpha \in \mathcal{S}'';\ j_\alpha = k,\ y_\alpha + \lambda_\alpha t = y_k^\theta(t) \ \text{ for some } t \in [0,\tau]\Big\}.$$

When a k-compression wave of strength $|\sigma_\alpha|$ impinges on the big k-shock of strength $|\sigma_k|$, an interaction of magnitude $|\sigma_k \sigma_\alpha| >> |\sigma_\alpha|^2$ takes place, and the functional Q decreases accordingly. This yields an estimate of the form

$$\big\|(v,\xi)(\tau)\big\|^{\heartsuit}_{w(\tau)} \le \big\|(v,\xi)0)\big\|^{\heartsuit}_{w(0)} + \varepsilon' - \kappa_1\kappa_2|\bar\sigma| \cdot \sum_{\alpha\in\mathcal{S}''_k} |\sigma_\alpha|^2. \tag{8.55}$$

Moreover, for $\theta \notin \Theta'$, when we replace w^θ by $\bar{w}^\theta$ the norm of the corresponding tangent vector satisfies the estimate

$$\big\|(\bar v,\bar\xi)\big\|^{\diamondsuit}_{\bar w} \le \big\|(\bar v,\bar\xi)\big\|^{\heartsuit}_{\bar w} \le \big\|(v,\xi)\big\|^{\heartsuit}_{w} - \kappa_1\kappa_2|\bar\sigma| \cdot \sum_{\alpha\in\mathcal{S}''\setminus\mathcal{S}''_k} |\sigma_\alpha|^2. \tag{8.56}$$

Together, (8.39), (8.55)-(8.56) and (8.33) yield

$$\begin{aligned}
\|\bar\gamma_\tau\|_\star &\le \|\bar\gamma_\tau\|_\diamondsuit + \varepsilon' \\
&\le \|\bar\gamma_0\|_\diamondsuit + 2\varepsilon' - \kappa_1\kappa_2(b-a-\varepsilon_0)|\bar\sigma| \cdot \sum_{\alpha\in\mathcal{S}''} |\sigma_\alpha|^2 + C\varepsilon_0 \\
&\le \|\gamma_0\|_\star + 2\varepsilon' + C'\varepsilon_0
\end{aligned}$$

for some constants C, C'. Since $\varepsilon', \varepsilon_0 > 0$ can be taken arbitrarily small, the result is proved.

7. The previous arguments yield a proof of Lemma 2.11 under the additional assumption that, for all $\theta \in \Theta$, in (8.11) we have $\hat{u}^\theta = \bar{u}^\theta$, i.e. that all initial data are constant outside the interval $[-4\rho,\ 4\rho]$.
To cover the general case, we first construct a path of approximate solutions w^θ on the trapezoid Γ, using the same procedures as before. Then we consider the path $\hat\gamma\colon \theta \mapsto \hat{w}^\theta$, where

$$\hat{w}^\theta(x) = \begin{cases} w^\theta(0,x) & \text{if } |x| > 4\rho, \\ w^\theta(T,x) & \text{if } |x| < 4\rho - T, \\ w^\theta(4\rho - |x|,\ x) & \text{if } |x| \in [4\rho - T,\ 4\rho]. \end{cases}$$

We then extend each w^θ on the outer region

$$\Gamma' \doteq \big\{(t,x);\ t \in [0,T],\ |x| > 4\rho - t\big\},$$

letting w^θ be an approximate ε-solution constructed by wave-front tracking, with initial data assigned on the space-like curve

$$t = \Lambda(x) = \begin{cases} 0 & \text{if } |x| > 4\rho, \\ 4\rho - |x| & \text{if } |x| \in [4\rho - T,\ 4\rho], \\ T & \text{if } |x| < 4\rho - T. \end{cases}$$

Observe that these extensions on Γ' are independent of θ. Indeed, all functions $w^\theta(T,\cdot)$ coincide for $|x| \ge 2\rho$. In particular, the tangent vectors (v^θ, ξ^θ) for the two paths $\hat\gamma\colon \theta \mapsto \hat{w}^\theta$ and $\gamma_T\colon \theta \mapsto w^\theta(T,\cdot)$ are supported inside the interval $[-\rho -$

$T,\ \rho+T]$ and coincide. Since the weights W_i^u in (8.4) can only decrease as a result of wave interactions, we conclude

$$\|\gamma_T\|_\star \leq \|\hat{\gamma}\|_\star \leq \|\gamma_0\|_\star .$$

This completes the proof of Lemma 2.11 in Case 2, i.e. when all functions $\bar{u}^\theta$ contain exactly one big shock inside the trapezoid Γ.

Construction of approximate solutions, Case 3

In the remainder of this section we describe the construction of piecewise Lipschitz approximate ε-solutions in Case 3, on a domain of functions containing two large approaching shocks. Since the flow of our semigroup S can be approximated by a cyclical concatenation of n distinct flows as in (8.12), it suffices to describe how to construct approximate trajectories for each semigroup S^h, $h = 1, \ldots, n$. We will also introduce a weighted norm $\|(v,\xi)\|_u^\diamond$ for tangent vectors, which is non-increasing along piecewise Lipschitz approximate solutions. This weighted norm incorporates some small terms due to the presence of shock layers with some width $\delta_2 > 0$ around the big shocks. As δ_2 approaches zero, for all piecewise Lipschitz u and $(v,\xi) \in T_u$ we will have the convergence

$$\lim_{\delta_2 \to 0+} \|(v,\xi)\|_u^\diamond = \|(v,\xi)\|_u^\star, \tag{8.57}$$

where $\|\cdot\|^\star$ is the norm at (8.3)–(8.6). The contractivity of the semigroup S^h w.r.t. the distance $d_\star$ is proved by the same arguments as in Case 2.

The domain of S^h has the form $\mathcal{D} \doteq \mathcal{D}^1 \cup \mathcal{D}^2$. There exists some constant states u^-, u^+ such that

$$\lim_{x\to-\infty} u(x) = u^-, \qquad \lim_{x\to\infty} u(x) = u^+$$

for all $u \in \mathcal{D}$. Moreover, each function $u \in \mathcal{D}^1$ contains two large approaching shocks, say of the k_1, k_2-characteristic families (with $k_1 \geq k_2$), located at points $y_1 < y_2$. These shocks have strength $|\sigma_1|, |\sigma_2| > \varepsilon^9$, and the total strength of all other waves in u is $< \varepsilon^{20}$. On the other hand, $\mathcal{D}^2 \doteq \mathcal{D}_{(u^-,u^+)}$ as in (2.8). If $u(0,\cdot) \in \mathcal{D}^1$, the solution u will remain inside $\mathcal{D}^1$ up to some time τ when the two large shocks interact, then it will evolve inside $\mathcal{D}^2$.
Let $u^\dagger \doteq \bar{u}\big((y_1+y_2)/2\big)$ be a middle state for some fixed function $\bar{u} \in \mathcal{D}^1$. By possibly shrinking the domain, since the total amount of small waves in any $u \in \mathcal{D}^1$ is $< \varepsilon^{20}$, we can assume

$$\max\Big\{\big|u(y_1-) - u^-\big|,\ \big|u(y_1+) - u^\dagger\big|,\ \big|u(y_2-) - u^\dagger\big|,\ \big|u(y_2+) - u^+\big|\Big\} \leq C\varepsilon^{20} \tag{8.58}$$

for every $u \in \mathcal{D}^1$. An approximate flow for the semigroup S^h will be constructed separately on the two domains $\mathcal{D}^1$ and $\mathcal{D}^2$.

We begin with the flow on $\mathcal{D}^1$. Introduce the following constant speeds. If $k_1 > k_2$, define

$$\lambda_j^* \doteq \lambda_j(u^-) \qquad \text{if } j \neq k_1, k_2,$$

$$\lambda_{k_1}^* \doteq \lambda_{k_1}(u^-), \quad \lambda_{k_1}^{**} \doteq \lambda_{k_1}(u^\dagger), \qquad \lambda_{k_2}^* \doteq \lambda_{k_2}(u^\dagger), \quad \lambda_{k_2}^{**} \doteq \lambda_{k_2}(u^+).$$

In the case $k_1 = k_2$ define

$$\lambda_j^* = \lambda_j^{**} \doteq \lambda_j(u^-) \quad \text{if } j \neq k_1,$$
$$\lambda_{k_1}^* \doteq \lambda_{k_1}(u^-), \quad \lambda_{k_1}^{**} \doteq \lambda_{k_1}(u^+), \quad \lambda_{k_1}^\dagger \doteq \lambda_{k_1}(u^\dagger).$$

For a fixed $h \in \{1, \ldots, n\}$ and a given $\delta_2 > 0$, we now introduce a system of equations which define our approximate solutions on the domain $\mathcal{D}^1$. In this system, all characteristic fields $j \neq h$ are linearly degenerate. On the other hand, the h-th eigenvalues are constant inside the two shock layers

$$[y_1^*, y_1^{**}] \doteq [y_1 - \delta_2,\ y_1 + \delta_2], \qquad [y_2^*, y_2^{**}] \doteq [y_2 - \delta_2,\ y_2 + \delta_2], \tag{8.59}$$

and genuinely nonlinear outside. More precisely, we consider the quasilinear hyperbolic system

$$u_t + A_h(x, u)u_x = 0, \tag{8.60}$$

where $A(x, u)$ is the matrix with the same eigenvectors as $A(u) = DF(u)$, but whose eigenvalues $\lambda_j^{(h)}$ are defined as follows. If $j \neq h$, then

$$\lambda_j^{(h)} \doteq \begin{cases} \lambda_j^* & \text{if } j \neq k_1, k_2, \\ \lambda_{k_1}^* & \text{if } j = k_1,\ x < y_1, \\ \lambda_{k_1}^{**} & \text{if } j = k_1 \neq k_2,\ x > y_1, \\ \lambda_{k_2}^* & \text{if } j = k_2 \neq k_1,\ x < y_2, \\ \lambda_{k_2}^{**} & \text{if } j = k_2,\ x > y_2, \\ \lambda_{k_1}^\dagger & \text{if } j = k_1 = k_2,\ y_1 < x < y_2. \end{cases}$$

If $h \notin \{k_1, k_2\}$, then

$$\lambda_h^{(h)} \doteq \lambda_h^* + n\big(\lambda_h(u) - \lambda_h^*\big).$$

If $h = k_1 \neq k_2$, then

$$\lambda_h^{(h)} \doteq \begin{cases} \lambda_{k_1}^* + n\big(\lambda_{k_1}(u) - \lambda_{k_1}^*\big) & \text{if } x < y_1^*, \\ \lambda_{k_1}^{**} + n\big(\lambda_{k_1}(u) - \lambda_{k_1}^{**}\big) & \text{if } x > y_1^{**}, \\ \lambda_{k_1}^* & \text{if } x \in [y_1^*, y_1[, \\ \lambda_{k_1}^{**} & \text{if } x \in]y_1, y_1^{**}]. \end{cases}$$

The definition of $\lambda_h^{(h)}$ in the case $h = k_2 \neq k_1$ is analogous. Finally, if $h = k_1 = k_2$ we set

$$\lambda_h^{(h)} \doteq \begin{cases} \lambda_h^* + n(\lambda_h(u) - \lambda_h^*) & \text{if } x < y_1, \\ \lambda_h^{**} + n(\lambda_h(u) - \lambda_h^{**}) & \text{if } x > y_2, \\ \lambda_h^\dagger + n(\lambda_h(u) - \lambda_h^\dagger) & \text{if } y_2 - y_1 > 2\delta_2 \text{ and } x \in]y_1^{**}, y_2^*[, \\ \lambda_h^* & \text{if } x \in [y_1^*, y_1[, \\ \lambda_h^{**} & \text{if } x \in]y_2, y_2^{**}], \\ \lambda_h^\dagger & \text{if } y_2 - y_1 > 2\delta_2 \text{ and } x \in]y_1, y_1^{**}[\cup[y_2^*, y_2[, \\ & \text{or if } y_2 - y_1 \leq 2\delta_2 \text{ and } x \in]y_1, y_2[. \end{cases}$$

Within the domain $\mathcal{D}^1$, an approximate ε-solution $u = u(t, x)$ is defined by the following requirements:

- At each time t, the function $u(t,\cdot)$ is piecewise Lipschitz, with two large shocks at points $y_1 < y_2$, and possibly other small h-shocks at points y_β, $\beta \in \mathcal{S}'$.
- The quasilinear equations (8.60) hold a.e. outside the shocks.
- The two large shocks satisfy the ε-Rankine Hugoniot equations (2.3)-(2.4).
- At each small h-shock y_β, the left and right state are related by (3.3). The speed $\dot{y}_\beta$ of the shock is determined by (3.9).

This determines the evolution up to the time τ where the two large shocks interact. To preserve the piecewise Lipschitz continuity of approximate solutions, restarting procedures are performed exactly as in Section 4. Notice that the definition of the characteristic speeds is given in such a way that when the two shock layers merge together we can replace them with a single shock layer $[y_1^*, y_2^{**}]$.

After the interaction time, since $\mathcal{D}^2 = \mathcal{D}_{(u^-,u^+)}$ is a special case of the domains considered at (2.8), the construction of approximate solutions is much the same as in Sections 3-4. The only difference is that now, immediately after the interaction time, the various shock layers around the big shocks emerging from the interaction are not well separated. Therefore, some care must be taken in defining the characteristic speeds and the weighted norm of tangent vectors in a neighborhood of the interaction point $(\tau, \bar{x})$.

Call $\omega_0 = u^-, \omega_1, \dots, \omega_n = u^+$ the constant states in the ε-solution of the Riemann problem with data (u^-, u^+). Let $\{y_k\,;\ k \in \mathcal{S}\}$ be the new set of large shocks. Define

$$y^*(t) \doteq \bar{x} - \delta_2 + t - \tau, \qquad\qquad y^{**}(t) \doteq \bar{x} + \delta_2 - t + \tau.$$

The characteristic speeds λ_i^*, λ_i^{**} are now defined as in (3.1), while the $\lambda_i^{(h)}$ are defined as in (3.4) for $i \neq h$. Concerning the genuinely nonlinear eigenvalues $\lambda_h^{(h)}$, on the time interval $[\tau,\ \tau + \delta_2/3]$ immediately after the interaction, we set

$$\lambda_h^{(h)} \doteq \begin{cases} \lambda_h^* & \text{if } h \in \mathcal{S},\ x \in \big[y^*(t),\ y_h(t)\big[\,, \\ \lambda_h^{**} & \text{if } h \in \mathcal{S},\ x \in \big]y_h(t), y^{**}(t)\big], \\ \lambda_h^* & \text{if } h \notin \mathcal{S},\ x \in \big[y^*(t),\ y^{**}(t)\big], \\ \lambda_h^* + n(\lambda_h(u) - \lambda_h^*) & \text{if } x < y^*(t), \\ \lambda_h^{**} + n(\lambda_h(u) - \lambda_h^{**}) & \text{if } x > y^{**}(t). \end{cases}$$

Due to the strict hyperbolicity, there exists a width $\delta_2' \in]0, \delta_2[$ such that the intervals

$$[y_i^*, y_i^{**}] \doteq [y_i - \delta_2',\ y_i + \delta_2'], \qquad\qquad i \in \mathcal{S}$$

are pairwise disjoint at time $t = \tau + \delta_2/3$. We can thus replace the shock layer $[y^*, y^{**}]$ with $|\mathcal{S}|$ new shock layers of width δ_2' around the big shocks, and define $\lambda_h^{(h)}$ as in (3.6) or (3.7). Since $u(\tau + \delta_2/3) \in \mathcal{D}_{(u^-,u^+)}$, a piecewise Lipschitz approximate solution can now be constructed for all $t \in [\tau + \delta_2/3,\ \infty[$, as in Sections 3-4.

In turn, letting the width of the shock layers $\delta_2, \delta_2' \to 0+$, in the limit we obtain the flow of the semigroup S^h. The proof of the contractivity of S^h w.r.t. the distance $d_\star$ is obtained by first showing that a suitable weighted norm $\big\|(v,\xi)\big\|_u^\diamond$ for tangent vectors decreases along our piecewise Lipschitz approximate solutions, and then arguing as in Case 2, exploiting the convergence (8.57) as $\delta_2 \to 0$.

In the present case, the analysis is somewhat longer, since we need to distinguish four evolution phases (Fig. 8.1):

(a) Before the interaction, when the two shock layers are still disjoint, with $u(t) \in \mathcal{D}^1$ and $y_2 - y_1 > 2\delta_2$.
(b) Slightly before the interaction, when the two shock layers have already merged together, with $u(t) \in \mathcal{D}^1$ and $y_2 - y_1 \leq 2\delta_2$.
(c) Immediately after the interaction, when all big shocks still remain inside a unique shock layer, with $u(t) \in \mathcal{D}^2$, $t \in [\tau, \tau + \delta_2/3]$.
(d) After the interaction, when the shock layers around the large shocks are mutually disjoint, with $u(t) \in \mathcal{D}^2$, $t > \tau + \delta_2/3$.

We now define a weighted norm $\|\cdot\|^{\diamondsuit}$ in each of these four phases. Let $u \in \mathcal{D}^1 \cup \mathcal{D}^2$ be piecewise Lipschitz, with big shocks at points y_α, $\alpha \in \mathcal{S}$, plus several small shocks at points y_β, $\beta \in \mathcal{S}'$. In analogy with (8.26), we set

$$\begin{aligned} \|(v,\xi)\|_u^{\diamondsuit} \doteq \sum_{i=1}^{n} \int_{-\infty}^{\infty} \widetilde{W}_i^u(x)\big|v_i(x)\big|dx + \sum_{\alpha\in\mathcal{S}\cup\mathcal{S}'} \widetilde{W}_{k_\alpha}^u|\sigma_\alpha||\xi_\alpha| \\ - \varepsilon^{19}\int_I \big|v_h(x)\big|\,dx, \end{aligned} \tag{8.61}$$

where $\widetilde{W}_i^u$ is defined as in (8.27). The terms $\widetilde{R}_i^u$ are now defined by

$$\widetilde{R}_i^u(x) \doteq \left[\sum_{j\leq i}\int_x^{\infty} + \sum_{j\geq i}\int_{-\infty}^{x}\right]\big|u_x^j(y)\big|\,dy + \left[\sum_{\substack{\alpha\in\mathcal{S}\cup\mathcal{S}'\\ k_\alpha\leq i,\ y_\alpha>x}} + \sum_{\substack{\alpha\in\mathcal{S}\cup\mathcal{S}'\\ k_\alpha\geq i,\ y_\alpha<x}}\right]|\sigma_\alpha| + P_i(x).$$

In the case $u \in \mathcal{D}^1$, the weights assigned to the big shocks at y_l, $l = 1, 2$ are defined by

$$\begin{aligned} \widetilde{R}_{k_l}^u(y_l) \doteq& \left[\sum_{j\leq k_l}\int_{y_l}^{\infty} + \sum_{j\geq k_l}\int_{-\infty}^{y_l}\right]\big|u_x^j(y)\big|\,dy \\ &+ \left[\sum_{\substack{\alpha\in\mathcal{S}'\\ k_\alpha\leq k_l,\ y_\alpha>y_l}} + \sum_{\substack{\alpha\in\mathcal{S}'\\ k_\alpha\geq k_l,\ y_\alpha<y_l}}\right]|\sigma_\alpha| + \varepsilon\hat{P}_{k_l}(y_l), \end{aligned}$$

while, if $u \in \mathcal{D}^2$, $k \in \mathcal{S}$,

$$\begin{aligned} \widetilde{R}_k^u(y_k) \doteq& \left[\sum_{j\leq k}\int_{y_k}^{\infty} + \sum_{j\geq k}\int_{-\infty}^{y_k}\right]\big|u_x^j(y)\big|\,dy \\ &+ \left[\sum_{\substack{\alpha\in\mathcal{S}'\\ k_\alpha\leq k,\ y_\alpha>y_k}} + \sum_{\substack{\alpha\in\mathcal{S}'\\ k_\alpha\geq k,\ y_\alpha<y_k}}\right]|\sigma_\alpha| + \varepsilon\hat{P}_k(y_k). \end{aligned}$$

The terms P_i, $\hat{P}_i$ and the domain $I \subset \mathbb{R}$ of the last integral in (8.61) are defined in different ways, according to the four cases (a)–(d) considered above.

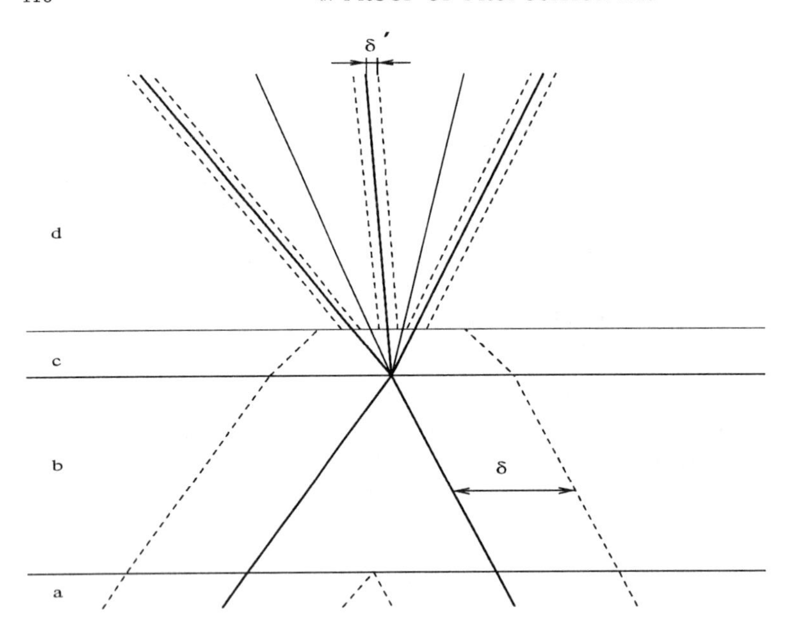

FIGURE 8.1

Case (a). $I = [y_1', y_1''] \cup [y_2', y_2'']$, where

$$y_l' \doteq \begin{cases} y_l^*, & \text{if } h \geq k_l, \\ y_l, & \text{if } h < k_l, \end{cases} \qquad y_l'' \doteq \begin{cases} y_l, & \text{if } h > k_l, \\ y_l^{**}, & \text{if } h \leq k_l, \end{cases} \qquad l \in \{1, 2\},$$

$$P_i(x) \doteq \begin{cases} 0, & \text{if } x \notin [y_1^*, y_1^{**}] \cup [y_2^*, y_2^{**}], \\ -\varepsilon^{19}, & \text{if } x \in [y_l^*, y_l[,\ i \geq k_l, \text{ or} \\ & \text{if } x \in]y_l, y_l^{**}],\ i \leq k_l, \text{ for some } l \in \{1,2\}, \\ \varepsilon^{19}, & \text{if } x \in [y_l^*, y_l[,\ i < k_l, \text{ or} \\ & \text{if } x \in]y_l, y_l^{**}],\ i > k_l, \text{ for some } l \in \{1,2\}, \end{cases}$$

and, for $l = 1, 2$,

$$\begin{aligned} \hat{P}_{k_l}(y_l) \doteq & \int_{y_l^*}^{y_l} \Big(\sum_{j<k_l} |u_x^j(x)| - \sum_{j \geq k_l} |u_x^j(x)| \Big) dx \\ & + \int_{y_l}^{y_l^{**}} \Big(\sum_{j>k_l} |u_x^j(x)| - \sum_{j \leq k_l} |u_x^j(x)| \Big) dx \\ & + \sum_{\substack{\beta \in S' \\ y_\beta \in [y_l^*, y_l^{**}] \setminus [y_l', y_l'']}} |\sigma_\beta| - \sum_{\substack{\beta \in S' \\ y_\beta \in [y_l', y_l'']}} |\sigma_\beta|. \end{aligned}$$

Case (b). $I = [y', y'']$, where

$$y' \doteq \begin{cases} y_1^*, & \text{if } h \geq k_1, \\ y_1, & \text{if } h < k_1, \end{cases} \qquad y'' \doteq \begin{cases} y_2^{**}, & \text{if } h \leq k_2, \\ y_2, & \text{if } h > k_2, \end{cases}$$

$$P_i(x) \doteq \begin{cases} 0, & \text{if } x < y_1^* \text{ or } x > y_2^{**}, \\ -\varepsilon^{19}, & \text{if } x \in]y_1, y_2[, \text{ or} \\ & \text{if } x \in [y_1^*, y_1[, \text{ and } i \geq k_1, \text{ or} \\ & \text{if } x \in]y_2, y_2^{**}], \text{ and } i \leq k_2, \\ \varepsilon^{19}, & \text{if } x \in [y_1^*, y_1[, \text{ and } i < k_1, \text{ or} \\ & \text{if } x \in]y_2, y_2^{**}], \text{ and } i > k_2, \end{cases}$$

and if $h \neq k_1$

$$\hat{P}_{k_1}(y_1) \doteq \int_{y_1^*}^{y_1} \left(\sum_{j<k_1} |u_x^j(x)| - \sum_{j \geq k_1} |u_x^j(x)| \right) dx - \int_{y_1}^{y_2^{**}} \sum_{j \leq k_1} |u_x^j(x)| \, dx +$$
$$+ \operatorname{sgn}(h - k_1) \left(\sum_{\substack{\beta \in \mathcal{S}' \\ y_\beta \in]y_1, y_2^{**}]}} |\sigma_\beta| - \sum_{\substack{\beta \in \mathcal{S}' \\ y_\beta \in [y_1^*, y_1[}} |\sigma_\beta| \right),$$

while if $h = k_1$ the last term is replaced by

$$- \sum_{\substack{\beta \in \mathcal{S}' \\ y_\beta \in [y_1^*, y_2^{**}]}} |\sigma_\beta|,$$

and if $h \neq k_2$

$$\hat{P}_{k_2}(y_2) \doteq - \int_{y_1^*}^{y_2} \sum_{j \geq k_2} |u_x^j(x)| \, dx + \int_{y_2}^{y_2^{**}} \left(\sum_{j>k_2} |u_x^j(x)| - \sum_{j \leq k_2} |u_x^j(x)| \right) dx +$$
$$+ \operatorname{sgn}(h - k_2) \left(\sum_{\substack{\beta \in \mathcal{S}' \\ y_\beta \in]y_2, y_2^{**}]}} |\sigma_\beta| - \sum_{\substack{\beta \in \mathcal{S}' \\ y_\beta \in [y_1^*, y_2[}} |\sigma_\beta| \right),$$

while if $h = k_2$ the last term is replaced by

$$- \sum_{\substack{\beta \in \mathcal{S}' \\ y_\beta \in [y_1^*, y_2^{**}]}} |\sigma_\beta|.$$

Case (c). $I = \emptyset$,

$$P_i(x) \doteq \begin{cases} \varepsilon^{19}, & \text{if } x < y^* \text{ or } x > y^{**}, \\ 2\varepsilon^{19}, & \text{if } x \in [y^*, y^{**}], \end{cases}$$

and, for $k \in \mathcal{S}$,

$$\hat{P}_k(\hat{y}_k) \doteq \int_{-\infty}^{+\infty} \sum_j |u_x^j(x)| \, dx + \int_{y^*}^{y^{**}} \sum_j |u_x^j(x)| \, dx.$$

Case (d). $I = \bigcup_{k \in \mathcal{S}} [y_k', y_k'']$, where

$$y_k' \doteq \begin{cases} y_k^*, & \text{if } h \geq k, \\ y_k, & \text{if } h < k, \end{cases} \qquad y_k'' \doteq \begin{cases} y_k, & \text{if } h > k, \\ y_k^{**}, & \text{if } h \leq k, \end{cases} \qquad k \in \mathcal{S},$$

$$P_i(x) \doteq \begin{cases} -\varepsilon^{19}, & \text{if } x \in [y_k^*, y_k[,\ i \geq k, \text{ or} \\ & \text{if } x \in]y_k, y_k^{**}],\ i \leq k, \text{ for some } k \in \mathcal{S}, \\ \varepsilon^{19}, & \text{if } x \in [y_k^*, y_k[,\ i < k, \text{ or} \\ & \text{if } x \in]y_k, y_k^{**}],\ i > k, \text{ for some } k \in \mathcal{S}, \\ 0, & \text{otherwise,} \end{cases}$$

and, for $k \in \mathcal{S}$, if $h \neq k$

$$\begin{aligned} \hat{P}_k(y_k) \doteq & \int_{y_k^*}^{y_k} \Big(\sum_{j<k} |u_x^j(x)| - \sum_{j \geq k} |u_x^j(x)|\Big) dx \\ & + \int_{y_k}^{y_k^{**}} \Big(\sum_{j>k} |u_x^j(x)| - \sum_{j \leq k} |u_x^j(x)|\Big) dx \\ & + \operatorname{sgn}(h-k) \left(\sum_{\substack{\beta \in \mathcal{S}' \\ y_\beta \in]y_k, y_k^{**}]}} |\sigma_\beta| - \sum_{\substack{\beta \in \mathcal{S}' \\ y_\beta \in [y_k^*, y_k[}} |\sigma_\beta| \right), \end{aligned}$$

while if $h = k$ the last term is replaced by

$$- \sum_{\substack{\beta \in \mathcal{S}' \\ y_\beta \in [y_k^*, y_k^{**}]}} |\sigma_\beta|.$$

Let us denote by I^a, I^b, I^c and I^d the set I defined in the cases (a), (b), (c), (d) respectively, and the same for the P_i's and the $\hat{P}_i$'s. By the same arguments in Sections 3-4, one checks that the weighted norm $\|(v,\xi)\|_u^\diamond$ of any tangent vector is non-increasing along our piecewise Lipschitz approximate solutions. Moreover, at every restarting time, the increase in the weighted norm can be kept arbitrarily small by increasing the accuracy of the restarting procedure. It remains to check that, at each time of transition from one of the phases (a)–(d) to the next, these weighted norms do not increase.

Transition from phase (a) to phase (b).

Let $\bar{t}$ be the transition time from Case (a) to Case (b). It is easy to check that $I^a \subset I^b$, $P_i^b(x) - P_i^a(x) \leq 0$ and $\hat{P}_{k_l}^b(y_l) - \hat{P}_{k_l}^a(y_l) \leq 0$. Therefore,

$$\|(v,\xi)\|_{u(\bar{t}+)}^\diamond \leq \|(v,\xi)\|_{u(\bar{t}-)}^\diamond. \tag{8.62}$$

Transition from phase (c) to phase (d).

Let $\bar{t}$ denote the transition time. Since $I^c = \emptyset$, we clearly have $I^c \subset I^d$. Due to the splitting of the shock layers, we may have an increase in $\|v\|_{\mathbf{L}^1}$. More precisely, recalling that the strength of the waves outside the big shocks is bounded by ε^{20}, we get

$$\|v(\bar{t}+)\|_{\mathbf{L}^1} \leq (1 + C\varepsilon^{20})\|v(\bar{t}-)\|_{\mathbf{L}^1}.$$

On the other hand, from the definition of P_i and $\hat{P}_i$, it easily follows that

$$\widetilde{W}_i^{u(\bar{t}+)}(x) - \widetilde{W}_i^{u(\bar{t}-)}(x) \leq -\kappa_1 \varepsilon^{19}$$

so that

$$\sum_i \int_{-\infty}^{+\infty} \left[|v_i^+(x)| \widetilde{W}_i^{u(\bar{t}+)}(x) - |v_i^-(x)| \widetilde{W}_i^{u(\bar{t}-)}(x) \right] dx$$
$$\leq \sum_i \|v_i^-\|_{\mathbf{L}^1} (C\varepsilon^{20} - \kappa_1 \varepsilon^{19}) \leq 0,$$

for ε small enough. This again implies (8.62).

Transition from phase (b) to phase (c).

We recall that $\bar{x}$ denotes the point of interaction of the two big shocks at time τ. Define $u^\pm \doteq u(\tau, \bar{x}\pm)$. Let us denote by $\{\xi_j\}_{j\in\mathcal{S}}$ the shifts of the shocks generated by the interaction. We recall that, since no rarefaction waves are generated by the interaction, we have

$$\text{(8.63)} \qquad \begin{aligned} &\lim_{t\to\tau+} \int_{-\infty}^{+\infty} \left| u_x^i(t,x) - u_x^i(\tau-,x) \right| dx = 0, \\ &\lim_{t\to\tau+} \int_{-\infty}^{+\infty} \left| v_i(t,x) - v_i(\tau-,x) \right| dx = 0, \end{aligned} \qquad i = 1, \dots, n.$$

In order to compute the variation of the norm of a generalized tangent vector at the interaction time $t = \tau$, we can assume that only one wave is shifted at time $t < \tau$. Let us denote by $\xi \in \mathbb{R}$ the shift of this wave and by σ its strength. We shall distinguish the four cases (i)-(iv) as in CASE 2.

Due to the possible change in P_i, the terms $\widetilde{R}_i^u$ satisfy the estimate

$$\widetilde{R}_i^{u(\tau+)}(x) \leq \widetilde{R}_i^{u(\tau-)}(x) + 3\varepsilon^{19}.$$

On the other hand, the interaction potential decreases at time τ:

$$Q(\tau+) \leq Q(\tau-) - |\sigma_1\sigma_2|.$$

Since $|\sigma_1|, |\sigma_2| > \varepsilon^9$, we get for every $x \neq \bar{x}$, $i = 1, \dots, n$,

$$\text{(8.64)} \qquad \widetilde{W}_i^{u(\tau+)}(x) - \widetilde{W}_i^{u(\tau-)}(x) \leq 3\kappa_1\varepsilon^{19} - \kappa_1\kappa_2|\sigma_1\sigma_2| \leq -C\kappa_1|\sigma_1\sigma_2|,$$

for ε small. In a similar way, at the interaction point,

$$\text{(8.65)} \qquad \begin{aligned} &\widetilde{W}_i^{u(\tau+)}(\hat{y}_{k_1}) - \widetilde{W}_i^{u(\tau-)}(y_1) \leq 3\kappa_1\varepsilon\cdot\varepsilon^{20} - \kappa_1\kappa_2|\sigma_2| \leq -C\kappa_1|\sigma_2|, \\ &\widetilde{W}_i^{u(\tau+)}(\hat{y}_{k_2}) - \widetilde{W}_i^{u(\tau-)}(y_2) \leq -C\kappa_1|\sigma_1|. \end{aligned}$$

Let us first consider the case $k_1 \neq k_2$. In the case (i), we can assume that the shifted wave is the large k_1-shock, the other case being entirely similar. By (5.10) in [**B4**], the new shifts are given by

$$\xi_i = \frac{\lambda_i(\omega_{i-1}, \omega_i) - \dot{y}_2^-}{\dot{y}_1^- - \dot{y}_2^-}\xi, \qquad i \in \mathcal{S}.$$

From well-known estimates (see [**B4**]) we get

$$\text{(8.66)} \qquad |\xi_{k_1} - \xi| \leq C|\sigma_1\sigma_2|\,|\xi|, \quad |\xi_{k_2}| \leq C|\sigma_1\sigma_2|\,|\xi|, \quad |\xi_i| \leq C|\xi|,\ i \in \mathcal{S}.$$

From (8.61), (8.63), (8.64) and (8.65) we obtain

$$
\begin{aligned}
\|(v^+,\xi^+)\|^{\diamondsuit}_{u(\tau+)} - \|(v^-,\xi^-)\|^{\diamondsuit}_{u(\tau-)} &\\
\leq \sum_{i\in\mathcal{S}} |\sigma_i^+|\,|\xi_i|W_i^+(y_i) - |\sigma_1|\,|\xi|W_{k_1}^-(y_1) + &\\
+ \sum_{i=1}^{n}\int_{-\infty}^{+\infty} |v_i^-(x)|\left[W_i^+(x) - W_i^-(x)\right] dx + &\\
+ \varepsilon^{19}\int_{y^*}^{y^{**}} |v_h(x)|\,dx \leq & \qquad (8.67)\\
\leq nC|\xi|\,|\sigma_1\sigma_2|\,M_W^{u(\tau+)} + |\sigma_1|\,|\xi_{k_1} - \xi|\,W_{k_1}^+(y_{k_1}) + &\\
+ |\sigma_1|\,|\xi|\left[W_{k_1}^+(y_{k_1}) - W_{k_1}^-(y_1)\right] + &\\
+ (\varepsilon^{19} - C\kappa_1|\sigma_1\sigma_2|)\int_{y^*}^{y^{**}} |v_h(x)|\,dx \leq &\\
\leq nC|\xi|\,|\sigma_1\sigma_2|\,M_W^{u(\tau+)} - C\kappa_1|\sigma_1\sigma_2|\,|\xi| \leq &\\
\leq 0, &
\end{aligned}
$$

if κ_1 is large enough. The other cases (ii), (iii) and (iv) can be treated similarly.

Let us now consider what happens when $k_1 = k_2$. We shall treat in detail the case (i), since the other cases can be treated in the same way. Following [**B4**], the estimate (8.66) is replaced by

$$
|\xi_{k_1}| \leq \frac{|\sigma_1| + C|\sigma_1\sigma_2|}{|\sigma_1| + |\sigma_2|}|\xi|, \qquad |\xi_i| \leq \frac{C|\xi|}{|\sigma_1| + |\sigma_2|}, \qquad i\in\mathcal{S}, \tag{8.68}
$$

$$
\begin{aligned}
&|\sigma_{k_1}^+ - \sigma_1 - \sigma_2| \leq C|\sigma_1\sigma_2|(|\sigma_1| + |\sigma_2|),\\
&|\sigma_i^+| \leq C|\sigma_1\sigma_2|(|\sigma_1| + |\sigma_2|),
\end{aligned} \qquad i\in\mathcal{S}, \tag{8.69}
$$

while the estimates (8.64) and (8.65) remain valid. Using (8.64), (8.65), (8.68) and (8.69) we obtain

$$
\begin{aligned}
\sum_{i\in\mathcal{S}} |\sigma_i^+|\,|\xi_i|\,W_i^+(y_i) - |\sigma_1|\,|\xi|\,W_{k_1}^-(y_1) \leq & nC|\sigma_1\sigma_2|(|\sigma_1| + |\sigma_2|)\frac{C|\xi|}{|\sigma_1| + |\sigma_2|}M_W^{u(0+)}\\
& + \left|\,|\sigma_{k_1}^+|\,|\xi_{k_1}| - |\sigma_1|\,|\xi|\,\right| W_{k_1}^+(y_{k_1})\\
& + |\sigma_1|\,|\xi|\left[W_{k_1}^+(y_{k_1}) - W_{k_1}^-(y_1)\right]\\
\leq & - C_1|\sigma_1\sigma_2|.
\end{aligned}
$$

The same arguments used in (8.67) now yield

$$
\|(v^+,\xi^+)\|^{\diamondsuit}_{u(\tau+)} - \|(v^-,\xi^-)\|^{\diamondsuit}_{u(\tau-)} \leq 0.
$$

CHAPTER 9

Proof of Proposition 2.15

As in Section 2, consider a regular path $\gamma_0 : \theta \mapsto \bar{u}^\theta$ of initial data, defined for $\theta \in \Theta \doteq]a, b[$. Call Θ^* the set of all parameter values $\theta \in \Theta$ for which the corresponding solution u^θ is structurally unstable. If $\bar{\theta}$ is an isolated point of Θ^*, the conclusion of Proposition 2.10 is clear. We thus study the case where $\bar{\theta}$ is a limit point of Θ^*. Let us sketch the main ingredients of the proof.

1. By induction on the integer m, we can assume that the weighted length of a path

$$\theta \mapsto \gamma_t(\theta) \doteq u^\theta(t,\cdot) = S^\varepsilon_{t-\tau} u^\theta(\tau,\cdot) \qquad t \in [\tau, \infty[\tag{9.1}$$

is non-increasing provided that

$$\sup_\theta Q\big(u^\theta(\tau)\big) \le (m-1)\varepsilon^{20}. \tag{9.2}$$

It then suffices to prove that the same conclusion holds when

$$\sup_\theta Q\big(u^\theta(\tau)\big) \le m\varepsilon^{20}. \tag{9.3}$$

2. Let (9.3) hold, and assume that $u^{\bar{\theta}}$ is structurally unstable. From the definitions, it is clear that $u^{\bar{\theta}}$ can have only finitely many points of instability. Let $(t^{\bar{\theta}}, x^{\bar{\theta}})$ be the first such point (w.r.t. time). Arguing as in Section 7, we can choose $r, \rho > 0$ such that, setting

$$\tau \doteq t^{\bar{\theta}} - r, \qquad J \doteq [x^{\bar{\theta}} - 4r,\ x^{\bar{\theta}} + 4r], \qquad I_\rho \doteq [\bar{\theta} - \rho,\ \bar{\theta} + \rho], \tag{9.4}$$

all functions $u^\theta(\tau,\cdot)$, $\theta \in I_\rho$, have the same wave structure on the interval J.

3. We then consider a modified path $\tilde{\gamma}_\tau : \theta \mapsto \tilde{u}^\theta(\tau)$, having the form

$$\tilde{u}^\theta(\tau, x) = u^\theta\big(\tau,\ x - \varphi(x)\big), \tag{9.5}$$

where φ is a smooth scalar map. For any given $\varepsilon' > 0$, we will show that the map φ can be chosen with

$$\|\varphi\|_{C^3} < \varepsilon', \tag{9.6}$$

and such that, for all but finitely many values of θ, the solution

$$\tilde{u}^\theta(t,\cdot) \doteq S^\varepsilon_{t-\tau} \tilde{u}^\theta(\tau)$$

is structurally stable on some strip $[\tau,\ \tau^\theta] \times \mathbb{R}$ and satisfies

$$Q\big(\tilde{u}^\theta(\tau^\theta)\big) \le Q\big(u^\theta(\tau)\big) - \varepsilon^{20}. \tag{9.7}$$

Thanks to (9.7), the contractivity property can thus be extended to all times $t > \tau^\theta$ by the inductive assumption (9.2).

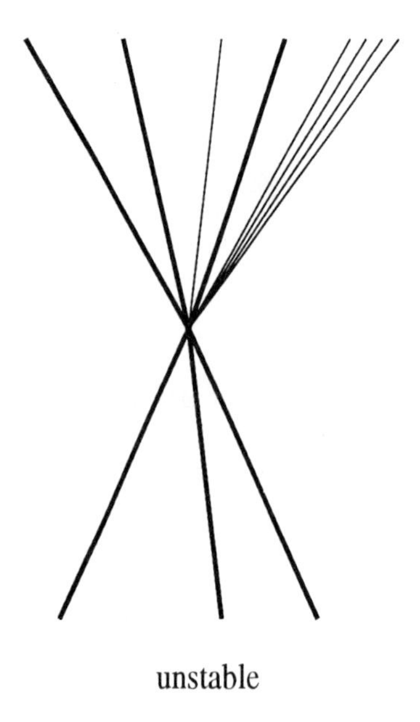

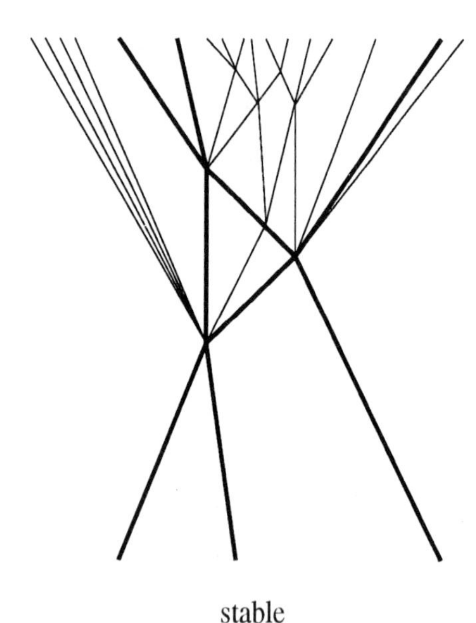

FIGURE 9.1

As in (2.23), define the rescaled functions

$$u^\eta(t,x) \doteq u^{\bar\theta}(t^{\bar\theta}+\eta t,\ x^{\bar\theta}+\eta x), \tag{9.8}$$

$$\omega(t,x) \doteq \lim_{\eta\to 0+} u^\eta(t,x). \tag{9.9}$$

To clarify the main ideas, we first give a proof in the typical case where, for $t<0$, ω contains exactly three incoming shocks, all of strength $\geq \varepsilon^2$, and no other waves (Fig. 9.1).
Under these assumptions, thanks to a structural stability argument (see [**B-LF2**]), one can find $r,\rho>0$ such that the following holds. Defining τ, J, I_ρ as in (9.4), all functions $u^\theta(\tau,\cdot)$, $\theta\in I_\rho$, have three large shocks of strength $> c\varepsilon$ on J, plus possibly other waves of total strength $< C\varepsilon^6$, for some constants $C,c>0$.

Call x_i^θ, $i=1,2,3$, the positions of the three large shocks in u^θ. By possibly shrinking the size of ρ, we can find some $r^*>0$ such that the intervals $[x_i^{\bar\theta}-3r^*,\ x_i^{\bar\theta}+3r^*]$ are mutually disjoint and

$$|x_i^\theta - x_i^{\bar\theta}|\leq r^* \qquad \theta\in I_\rho,\quad i=1,2,3. \tag{9.10}$$

Let $\psi\colon \mathbb{R}\mapsto[0,1]$ be a smooth function such that

$$\psi(s)=\begin{cases}1 & \text{if } |s|\leq 1,\\ 0 & \text{if } |s|\geq 2.\end{cases} \tag{9.11}$$

For every $\zeta=(\zeta_1,\zeta_2,\zeta_3)\in\mathbb{R}^3$, define

$$\varphi_\zeta(\theta,x)\doteq\sum_{i=1}^3 \zeta_i\,\psi\left(\frac{x-x_i^{\bar\theta}}{r^*}\right). \tag{9.12}$$

Now fix any $\theta \in I_\rho$ and call $u^{\theta,\zeta}$ the ε-solution of (1.1) with initial data

$$u^{\theta,\zeta}(\tau,x) = u^\theta\big(\tau, x - \varphi_\zeta(x)\big). \tag{9.13}$$

Call $x_i^{\theta,\zeta}$ the locations of the corresponding shocks. The next step of the proof will establish that, roughly speaking,

$$\frac{\partial}{\partial\zeta_j} x_i^{\theta,\zeta}(t) \;\approx\; \delta_{ij} \doteq \begin{cases} 1 & \text{if } i = j, \\ 0 & \text{if } i \neq j, \end{cases} \tag{9.14}$$

for all times $t \geq \tau$ before the first interaction between two large shocks.

Observe that the two sides of (9.14) coincide at time τ. Moreover, if only a minimal amount of interaction takes place during the interval $[\tau, t]$, then the shift rates of the shocks x_i (w.r.t. the parameter ζ) will remain almost unchanged. Based on these ideas, we now work out a rigorous argument. For notational simplicity, the dependence on θ will henceforth be omitted. Since the map $\zeta \mapsto x_i^\zeta(t)$ is Lipschitz continuous but possibly not differentiable, the estimate (9.14) must be reformulated in terms of Clarke's generalized gradient [**Cl**]. More precisely, for $i = 1,2,3$ and all ζ sufficiently close to the origin, we claim that

$$\sup_{j=1,2,3} |\xi_j - \delta_{ij}| \leq C\varepsilon^5, \tag{9.15}$$

for every vector $\xi = (\xi_1, \xi_2, \xi_3)$ contained in the generalized gradient of the scalar map $\zeta \mapsto x_i^\zeta(t)$.

To establish (9.15), we show that the same estimate holds uniformly for the generalized gradients of piecewise Lipschitz approximate solutions. Indeed, for any fixed $\theta \in I_\rho$, consider a piecewise Lipschitz approximation $w(\tau,\cdot)$ of $u^\theta(\tau,\cdot)$, containing three large shocks located at the same points x_i^θ as u^θ, plus other waves of total strength $< C\varepsilon^6$. Then the regular path

$$\vartheta \mapsto u^\vartheta(x) \doteq w\big(\tau,\; x - \vartheta\varphi_\zeta(x)\big) \tag{9.16}$$

generates a generalized a tangent vector $(v^\vartheta, \xi^\vartheta)$. Its continuous part is

$$v^\vartheta(\tau,x) = -w_x\big(\tau,\; x - \vartheta\varphi_\zeta(x)\big)\; \varphi_\zeta(x). \tag{9.17}$$

Moreover, denoting by ξ_i^ϑ, $i \in \mathcal{S} \doteq \{1,2,3\}$ the shifts of the big shocks at x_i^ϑ and by ξ_α^ϑ, $\alpha \in \mathcal{S}'$ the shifts of the small shocks at x_α^ϑ, from (9.10)–(9.13) it follows

$$\xi_i^\vartheta(\tau) = \zeta_i, \qquad \big|\xi_\alpha^\vartheta(\tau)\big| \leq \max_i |\zeta_i|. \tag{9.18}$$

As long as the large shocks remain separated, we can construct piecewise Lipschitz approximations, as in Sections 3-4, with restartings that do not change the positions of these three shocks. We claim that, as long as these shocks do not interact, for any such path of approximate solutions $w^\vartheta(t,\cdot)$, the tangent vector $\big(v^\vartheta(t), \xi^\vartheta(t)\big)$ satisfies

$$\big|\xi_i(t) - \xi_i(\tau)\big| \leq C\varepsilon^4 \big|\xi(\tau)\big|. \tag{9.19}$$

A proof of (9.19) will be obtained from (9.17)-(9.18), by showing that the tangent vector remains almost unchanged during the time interval $[\tau, t]$, since the amount of interaction is very small.

Call σ_i, σ_α the sizes of the shocks at x_i, $i \in \mathcal{S}$ and at x_α, $\alpha \in \mathcal{S}'$. Let these shock belong to the families k_i, k_α, respectively. Fix $i \in \mathcal{S}$. By the same type of arguments used in Section 3 for proving the decrease in the weighted norm of tangent vectors, as long as $\big|\xi_i(t)-\xi_i(\tau)\big| \le \big|\xi_i(\tau)\big|/2$, we have

$$\frac{d}{dt}\Bigg[W_{k_i}^{w(t)}\big(x_i(t)\big)\big|\sigma_i(t)\big|\big|\xi_i(t)-\xi_i(\tau)\big| + \sum_{\alpha\in\mathcal{S}'\cup\mathcal{S}\setminus\{i\}} W_{k_\alpha}^{w(t)}\big(x_\alpha(t)\big)\big|\sigma_\alpha(t)\big|\big|\xi_\alpha(t)\big| + \sum_{j=1}^{n}\int_{-\infty}^{+\infty} W_j^{w(t)}(x)\big|v_j(t,x)\big|dx\Bigg] \le 0. \tag{9.20}$$

Let us estimate the increase of ξ_i and ξ_α at times when an interaction between a big shock and a small shock takes place. In this case, denoting by $\xi_i^\pm$ and $\xi_\alpha^\pm$ the shifts before and after the interaction, one has

$$|\xi_i^+ - \xi_i^-| \le \frac{|\xi_\alpha^-\sigma_\alpha^-|}{|\sigma_i^-|}|\xi_i^-| \qquad |\xi_\alpha^+ - \xi_\alpha^-| \le \frac{|\xi_i^-\sigma_\alpha^-|}{|\sigma_i^-|}|\xi_\alpha^-|. \tag{9.21}$$

For interactions between two small shocks, the identities (3.72) hold.
Furthermore, as long as the three large shocks remain separated, one has

$$\sum_{i=1}^{3}\Big|W_{k_i}^{w(t)}\big(x_i(t)\big)\big|\sigma_i(t)\big|\big|\xi_i(t)\big| - W_{k_i}^{w(\tau)}\big(x_i(\tau)\big)\big|\sigma_i(t)\big|\big|\xi_i(\tau)\big|\Big| \le C\varepsilon^6|\zeta|, \tag{9.22}$$

$$\sum_{j=1}^{n}\bigg|\int_{-\infty}^{+\infty}\Big(W_j^{w(t)}(x)\big|v_j(t,x)\big| - W_j^{w(\tau)}(x)\big|v_j(\tau,x)\big|\Big)\,dx\bigg| \le C\varepsilon^6|\zeta|. \tag{9.23}$$

The estimates (3.72) and (9.20)–(9.23) together imply

$$\big|\xi_i(t)-\xi_i(\tau)\big| \le C\varepsilon^6\frac{\big|\xi_i(\tau)\big|}{\big|\sigma_i(\tau)\big|} \le C\varepsilon^5\big|\xi_i(\tau)\big|. \tag{9.24}$$

Since (9.24) is uniformly valid for all suitably accurate approximate solutions, combining (9.24) with (9.18) and passing to the limit we obtain (9.15).

We now consider solutions

$$u^{\theta,\zeta}(t) = S^\varepsilon_{t-\tau}u^{\theta,\zeta}(\tau) \tag{9.25}$$

obtained with the special choice $\zeta = (0,\zeta_2,0)$. By genuine nonlinearity, the speeds of the three large shocks are strictly separated, i.e.

$$\inf_t\big\{\dot x_1^{\theta,\zeta}(t) - \dot x_2^{\theta,\zeta}(t),\ \dot x_2^{\theta,\zeta}(t) - \dot x_3^{\theta,\zeta}(t)\big\} > c\varepsilon^2. \tag{9.26}$$

By (9.15) and (9.26), for each $\theta\in I_\rho$ there exists a unique value $\zeta_2^*(\theta)$ such that, taking $\zeta = (0,\zeta_2^*,0)$, the three large shocks in the corresponding solution (9.25) interact together at a single point. Moreover, the map $\theta\mapsto\zeta_2^*(\theta)$ is Lipschitz continuous. Applying the coarea formula (see [**E-G**]), we obtain

$$\int_{-\infty}^{+\infty}\mathcal{H}^0\{\theta;\ \zeta_2^*(\theta)=\sigma\}d\sigma = \int_{I_\rho}\left|\frac{\partial\zeta_2^*(\theta)}{\partial\theta}\right|d\theta < +\infty,$$

where $\mathcal{H}^0$ is the zero dimensional Hausdorff measure, i.e. the counting measure. It follows that $\mathcal{H}^0\{\theta;\ \zeta_2^*(\theta)=\sigma\} < +\infty$ for almost every σ. Observing that $\|\varphi_\zeta\|_{C^3}\le C|\zeta|$, for any $\varepsilon'>0$ we can now choose $\zeta=(0,\sigma,0)$ so small that (9.6) holds, and such that the set $\big\{\theta;\ \zeta_2^*(\theta)=\sigma\big\}$ is finite.

The previous analysis shows that the path $\theta \mapsto u^\theta$ can be uniformly approximated by a path $\theta \mapsto u^{\theta,\zeta}$ such that, for all but finitely many θ, the solution $u^{\theta,\zeta}$ is structurally stable. The conclusion thus follows from Corollary 2.14 and the lower semicontinuity of the weighted length.

The general case can be handled by similar techniques. By a structural stability argument (see [**B-LF2**]) we can choose $r, \rho > 0$ such that, defining τ, J, I_ρ as in (9.4), the following holds. There exist disjoint closed intervals $J_1, \ldots, J_m \subset J$ such that all functions $u^\theta(\tau,\cdot)$, $\theta \in I_\rho$ contain a similar "wave packet" inside each J_ℓ. More precisely, one of the two cases occurs:

(a) Inside J_ℓ, each $u^\theta(\tau)$ contains a k_ℓ-shock of strength $\geq \varepsilon^4$, plus other waves of total strength $< \varepsilon^6$.

(b) Inside J_ℓ, each $u^\theta(\tau)$ contains an amount $V_{k_\ell}^-$ of negative k_ℓ-waves, with

$$V_{k_\ell}^-(u(\tau);\ J_\ell) \in [\varepsilon^3,\ 2\varepsilon^3],$$

while the total strength of positive k_ℓ-waves and of all other j-waves, $j \neq k_\ell$ is $\leq \varepsilon^6$.

We then construct smooth functions $\psi_\ell : \mathbb{R} \mapsto [0,1]$ with disjoint compact supports, such that $\psi_\ell \equiv 1$ on a neighborhood of J_ℓ. Given $\zeta = (\zeta_1, \ldots, \zeta_m)$, we define the modified functions $u^{\theta,\zeta}(\tau, x)$ as in (9.13), with

$$\varphi_\zeta(x) \doteq \sum_{\ell=1}^m \zeta_\ell \psi_\ell(x). \tag{9.27}$$

Now fix $\theta \in I_\rho$ and a unit vector $\zeta \in \mathbb{R}^m$, and consider the one-parameter family of solutions

$$\tilde{u}^\vartheta(t,\cdot) \doteq S_{t-\tau}^\varepsilon u^{\theta,\vartheta\cdot\zeta}(\tau).$$

At the initial time τ, by construction each wave packet shifts with ϑ at rate $\xi_\ell(\tau) = \zeta_\ell$. As long as the interaction remains small, say

$$Q(\tilde{u}^\vartheta(\tau)) - Q(\tilde{u}^\vartheta(t)) \leq \varepsilon^{20},$$

in the solution $\tilde{u}^\vartheta$ we can still identify m wave packets whose barycenters shift with ϑ at rates $\xi_\ell(t) \approx \zeta_\ell$. By a genericity argument based on the coarea formula [**E-G**], for any $\varepsilon' > 0$ we can thus choose $\zeta \in \mathbb{R}^m$ so small that (9.6) holds, and such that, for all but finitely many $\theta \in I_\rho$, the corresponding function $\tilde{u}^{\theta,\zeta}$ in (9.25) is structurally stable up to some time τ^θ where at least two wave-packets have interacted so that (9.7) holds. The conclusion of Proposition 2.10 now follows from the inductive assumption at (9.2).

CHAPTER 10

Completion of the proof

At this final stage, for every $\varepsilon > 0$ our previous analysis has established the existence of a semigroup $S^\varepsilon : \mathcal{D}^\varepsilon \times [0, \infty[\mapsto \mathcal{D}^\varepsilon$ of ε-solutions of (1.1). All these semigroups are continuous with a uniform Lipschitz constant L, and all domains $\mathcal{D}^\varepsilon$ contain a set $\{u \in \mathbf{L}^1;\ \text{Tot.Var.}(u) < \eta_0\}$, with $\eta_0 > 0$ independent of ε. We can now choose $\delta_0 > 0$ small enough so that every function u in the corresponding set $\mathcal{D}$ in (1.3) has total variation $< \eta_0$. In particular, this choice implies $\mathcal{D} \subseteq \mathcal{D}^\varepsilon$ for every ε. We claim that, as $\varepsilon \to 0$, the limit

$$S_t \bar{u} \doteq \lim_{\varepsilon \to 0+} S_t^\varepsilon \bar{u} \tag{10.1}$$

exists for all $t \geq 0$ and all initial data $\bar{u} \in \mathcal{D}$, and that S is a Standard Riemann Semigroup generated by the system of conservation laws (1.1).

To prove the claim, let $\bar{u} \in \mathcal{D}$, $\delta > 0$ be given. Then there exists a piecewise constant approximate solution $w = w(t, x)$ of the Cauchy problem (1.1)-(1.2), constructed by wave-front tracking, with $w(t) \in \mathcal{D}$ for all $t \geq 0$, such that

(i) $\|w(0) - \bar{u}\|_{\mathbf{L}^1} < \delta$,
(ii) all rarefaction fronts have size $< \delta$,
(iii) at every time t, the total strength of non-physical wave-fronts is $< \delta$.

In the following, we indicate by $\mathcal{S}, \mathcal{R}$ and $\mathcal{NP}$ respectively the set of shock, rarefaction and non-physical wave-fronts of $w(t)$. Using the Lipschitz continuity of the semigroups S^ε and the definition of ε-solutions, we obtain

$$\begin{aligned}
&\|w(\tau) - S_\tau^\varepsilon \bar{u}\|_{\mathbf{L}^1} \leq \\
&\leq L \cdot \|w(0) - \bar{u}\|_{\mathbf{L}^1} + L \int_0^\tau \limsup_{h \to 0+} \frac{\|w(t+h) - S_h^\varepsilon w(t)\|_{\mathbf{L}^1}}{h}\, dt \\
&\leq L\delta + L\tau \cdot \left\{ C \cdot \sum_{\alpha \in \mathcal{S}} \Big(\min\{|\sigma_\alpha|,\ \varepsilon\} \Big)^2 + C \cdot \sum_{\alpha \in \mathcal{R}} |\sigma_\alpha|^2 + C \cdot \sum_{\alpha \in \mathcal{NP}} |\sigma_\alpha| \right\} \\
&\leq L\delta + LC\tau \Big\{ \varepsilon \cdot \sum_{\alpha \in \mathcal{S}} |\sigma_\alpha| + \delta \cdot \sum_{\alpha \in \mathcal{R}} |\sigma_\alpha| + \delta \Big\} \\
&\leq L\delta + LC\tau \big\{ (\varepsilon + \delta) C' \cdot \text{Tot.Var.}(\bar{u}) + \delta \big\}
\end{aligned} \tag{10.2}$$

for some constants C, C'. We use here the fact that the total strength of waves in $w(t)$ can be bounded in terms of the total variation of $\bar{u}$. Repeating the above estimate with the same approximate solution w but with a different semigroup $S^{\varepsilon'}$, we obtain

$$\|w(\tau) - S_\tau^{\varepsilon'} \bar{u}\|_{\mathbf{L}^1} \leq L\delta + LC\tau \big\{ (\varepsilon' + \delta) C' \cdot \text{Tot.Var.}(\bar{u}) + \delta \big\}. \tag{10.3}$$

Since $\delta > 0$ can be chosen arbitrarily small, from (10.2)-(10.3) it follows

$$\left\| S_\tau^\varepsilon \bar{u} - S_\tau^{\varepsilon'} \bar{u} \right\|_{\mathbf{L}^1} \leq LC'' \tau(\varepsilon + \varepsilon') \tag{10.4}$$

valid for some constant C'' and all $\bar{u} \in \mathcal{D}$, $\varepsilon, \varepsilon' > 0$. This proves that the limit in (10.1) is well defined. The uniform continuity of S and the semigroup property are clear. To show that S acts correctly on piecewise constant initial data, let $\bar{u} \in \mathcal{D}$ be piecewise constant, say with jumps at the points $x_1 < \ldots < x_N$. Define

$$\tau \doteq \min_{i=2,\ldots,N} \frac{|x_i - x_{i-1}|}{2}.$$

Since by assumption all wave speeds are < 1 in absolute value, for every $t \in [0, \tau]$ and $\varepsilon > 0$, the function $u^\varepsilon(t, \cdot) = S_t^\varepsilon \bar{u}$ is obtained by piecing together the ε-solutions of the N Riemann problems corresponding to the jumps in $\bar{u}$. As $\varepsilon \to 0$, by our definitions it is clear that each ε-solution tends to the exact solution of the corresponding Riemann problem. Hence, for $t \in [0, \tau]$, the function $S_t \bar{u}$ satisfies the condition (iii) in the statement of Theorem 1.1. This completes the proof, in the case where all characteristic fields are genuinely nonlinear.

The linearly degenerate case

We conclude this section by describing the minor changes needed in the case where one or more characteristic fields are linearly degenerate. In the proof of Theorem 2.3, the assumption of genuine nonlinearity is used on three occasions. Namely, it guarantees the following properties:
- The boundary layers around each big shock are crossed transversally by waves of all characteristic families.
- For approximate solutions constructed by wave-front tracking, the local interaction potential in a forward neighborhood of every point in the t-x plane is arbitrarily small, as stated in (6.4).
- The decay estimates (1.21) for positive waves hold. These estimates are essential for proving results on structural stability.

To fix the ideas, assume now that the k-th family in the system (1.1) is linearly degenerate. To handle this case, we shall approximate (1.1) with another system where all characteristic fields are genuinely nonlinear. More precisely, the definition of ε-solutions is modified as follows.

Choose a unit vector $\mathbf{e}_k$ such that

$$\langle \mathbf{e}_k, \, r_k(u) \rangle > 0$$

for all u. Introduce the modified characteristic speed

$$\tilde{\lambda}_k^\varepsilon(u) \doteq \lambda_k(u) + \varepsilon \langle \mathbf{e}_k, \, r_k(u) \rangle. \tag{10.5}$$

For each $\varepsilon > 0$, we then define S^ε as the semigroup which acts on piecewise constant initial data according to the following approximate Riemann solver.

Modified ε-Riemann Solver: Given u^-, u^+, determine intermediate states $\omega_0 = u^-, \ldots, \omega_n = u^+$ and wave sizes $\sigma_1, \ldots, \sigma_n$ such that

$$\omega_i = \Psi_i^\varepsilon(\sigma_i)(\omega_{i-1}) \qquad i \neq k, \tag{10.6}$$

$$\omega_k = R_k(\sigma_k)(\omega_{k-1}). \tag{10.7}$$

- If $\sigma_i \geq 0$ and $i \neq k$, the states ω_{i-1}, ω_i are connected by a centered rarefaction wave, as usual.
- If $\sigma_i < 0$ and $i \neq k$, the states ω_{i-1}, ω_i are connected by a single jump, travelling with the speed $\dot{y} \doteq \lambda_i^\varepsilon(\omega_{i-1}, \omega_i)$ defined at (2.4).
- If $\sigma_k > 0$, then ω_{k-1} and ω_k are connected by a centered rarefaction k-wave travelling with characteristic speed $\tilde{\lambda}_k^\varepsilon$. In the sector where $t\tilde{\lambda}_k^\varepsilon(\omega_{k-1}) < x < t\tilde{\lambda}_k^\varepsilon(\omega_k)$, the ε-solution u thus satisfies

$$\big\langle l_i(u),\ u_x \big\rangle = 0 \qquad i \neq k, \qquad\qquad u_t + \tilde{\lambda}_k^\varepsilon(u) u_x = 0.$$

- If $\sigma_k < 0$, then ω_{k-1} and ω_k are connected by a k-jump satisfying (10.7) and travelling with speed

$$\dot{y} = \frac{1}{|\sigma_k|} \int_{\sigma_k}^0 \tilde{\lambda}_k^\varepsilon\big(R_k(s)(u^-)\big)\, ds. \tag{10.8}$$

The construction of piecewise Lipschitz approximate solutions is performed as before, except that we no longer insert boundary layers around big k-shocks. Indeed, such boundary layers are not needed in the present situation, because by construction shock and rarefaction curves of the k-th family always coincide. The rest of the analysis is entirely similar.

CHAPTER 11

Appendix

Estimates (3.16), (3.17), (3.21)

Let $u^-, u^+ \in \Omega$. We shall denote by $E(u^-,u^+) \doteq (E_1,\dots,E_n)(u^-,u^+)$ the wave sizes corresponding to the solution of the Riemann problem (u^-,u^+), obtained by the ε-Riemann Solver described in Section 2. By definition, for a fixed $k_\alpha \in \{1,\dots,n\}$ we have

$$E_{k_\alpha}(u^-, \Psi^\varepsilon_{k_\alpha}(\sigma)(u^-)) = \sigma, \quad \forall\sigma,$$
$$E_{k_\alpha}(u^-, \Psi^\varepsilon_{i}(\sigma)(u^-)) = 0, \quad \forall\sigma,\ \forall i \neq k_\alpha.$$

Assume $u^+ = \Psi^\varepsilon_{k_\alpha}(\sigma_\alpha)(u^-)$. Differentiating w.r.t. σ the above relations one obtains

$$D_2 E_{k_\alpha}(u^-,u^+)\cdot (r_{k_\alpha}(u^+) + O(\sigma_\alpha^2)) = 1, \tag{11.1}$$
$$D_2 E_{k_\alpha}(u^-,u^-)\cdot r_i(u^-) = 0, \quad \forall i \neq k_\alpha. \tag{11.2}$$

From (11.2) we can deduce that

$$|DE_{k_\alpha}(u^-,u^+)\cdot(0,v)| \leq C|\sigma_\alpha|, \quad \forall |v|=1,\ \langle v, r_{k_\alpha}(u^+)\rangle = 0. \tag{11.3}$$

In the same way we can obtain the relation

$$|DE_{k_\alpha}(u^-,u^+)\cdot(v,0)| \leq C|\sigma_\alpha|, \quad \forall |v|=1,\ \langle v, r_{k_\alpha}(u^-)\rangle = 0. \tag{11.4}$$

Let us define $r_i^\pm \doteq r_i(u(y_\alpha \pm))$. From (11.1) we get

$$|DE_{k_\alpha}(u^-,u^+)(0, r^+_{k_\alpha}) - 1| \leq C|\sigma_\alpha|^2. \tag{11.5}$$

Furthermore, by a similar argument one can deduce that

$$|DE_{k_\alpha}(u^-,u^+)(r^-_{k_\alpha}, 0) + 1| \leq C|\sigma_\alpha|^2. \tag{11.6}$$

Since $\sigma_\alpha = E_{k_\alpha}(u^-,u^+)$, differentiating w.r.t. $w^+_{k_\alpha}$ one obtains

$$\frac{\partial \sigma_\alpha}{\partial w^+_{k_\alpha}} = DE_{k_\alpha}(u^-,u^+)\left(\sum_{j<k_\alpha} r_j^- \frac{\partial W^j}{\partial w^+_{k_\alpha}}, r^+_{k_\alpha} + \sum_{j>k_\alpha} r_j^+ \frac{\partial W^j}{\partial w^+_{k_\alpha}}\right) =$$
$$= DE_{k_\alpha}(u^-,u^+)(0, r^+_{k_\alpha}) + DE_{k_\alpha}(u^-,u^+)\left(\sum_{j<k_\alpha} r_j^- \frac{\partial W^j}{\partial w^+_{k_\alpha}}, \sum_{j>k_\alpha} r_j^+ \frac{\partial W^j}{\partial w^+_{k_\alpha}}\right).$$

From (3.15) and (11.5) it easily follows that

$$\left|\frac{\partial \sigma_\alpha}{\partial w^+_{k_\alpha}} - 1\right| \leq C|\sigma_\alpha|^2.$$

In a similar way, using (11.6), one can prove that

$$\left|\frac{\partial \sigma_\alpha}{\partial w^-_{k_\alpha}} + 1\right| \leq C|\sigma_\alpha|^2.$$

The simpler estimate (3.16) can be obtained from (11.3) and (11.4) observing that

$$\frac{\partial \sigma_\alpha}{\partial w_i^+} = DE_{k_\alpha}(u^-, u^+)\left(\sum_{j<k_\alpha} r_j^- \frac{\partial W^j}{\partial w_i^+}, r_i^+\right), \qquad i^+ \in \mathcal{I},\ i \neq k_\alpha,$$

$$\frac{\partial \sigma_\alpha}{\partial w_i^-} = DE_{k_\alpha}(u^-, u^+)\left(r_i^-, \sum_{j>k_\alpha} r_j^+ \frac{\partial W^j}{\partial w_i^-}\right), \qquad i^- \in \mathcal{I},\ i \neq k_\alpha.$$

Concerning (3.21), the first relation is an easy consequence of the fact that, for a small h-shock, the left and the right states lie on the same h-rarefaction curve. The second relation in (3.21) follows from the identities

$$DE_{k_\alpha}(u^-, u^+)(r^-_{k_\alpha}, 0) = -1, \quad DE_{k_\alpha}(u^-, u^+)(0, r^+_{k_\alpha}) = 1, \tag{11.7}$$

$$DE_{k_\alpha}(u^-, u^+)(r_i^-, 0) = DE_{k_\alpha}(u^-, u^+)(0, r_i^+) = 0, \quad \forall i \neq k_\alpha. \tag{11.8}$$

Estimates (3.18), (3.19), (3.20), (3.22) The estimate (3.18) is trivial. We now prove (3.20); (3.19) can be obtained by a similar argument. Let us assume $\alpha \in \mathcal{S}$. At the point y_α the ε-Rankine-Hugoniot conditions (2.3)-(2.5) hold with $\sigma = \sigma_\alpha$. More precisely, we have that

$$\dot{y}_\alpha = \varphi\left(\frac{\sigma_\alpha}{\varepsilon}\right)\lambda^s_{k_\alpha} + \left(1 - \varphi\left(\frac{\sigma_\alpha}{\varepsilon}\right)\right)\lambda^r_{k_\alpha}, \tag{11.9}$$

where $\lambda^s_{k_\alpha} = \lambda_{k_\alpha}(u^-, S_{k_\alpha}(\sigma_\alpha)(u^-))$ and

$$\lambda^r_{k_\alpha} = \frac{1}{|\sigma_\alpha|}\int_{\sigma_\alpha}^0 \lambda_{k_\alpha}(R_{k_\alpha}(s)(u^-))\, ds.$$

Observe now that

$$|\lambda^s_{k_\alpha} - \lambda^r_{k_\alpha}| \leq C|\sigma_\alpha|^2. \tag{11.10}$$

Indeed the two functions

$$\sigma \mapsto \frac{1}{|\sigma|}\int_\sigma^0 \lambda_{k_\alpha}(R_{k_\alpha}(s)(u^-))\, ds, \quad \sigma \mapsto \lambda_{k_\alpha}(u^-, S_{k_\alpha}(\sigma)(u^-)), \qquad \sigma \leq 0,$$

have a tangency of the first order at $\sigma = 0$. By differentiating (11.9) w.r.t. $w^+_{k_\alpha}$, using (11.10) and by the fact that $\varphi'(\sigma_\alpha/\varepsilon) = 0$ if $\sigma_\alpha \leq -4\varepsilon$ or $\sigma_\alpha \geq -3\varepsilon$, we obtain the estimate

$$\begin{aligned} &\left|\frac{\partial \dot{y}_\alpha}{\partial w^+_{k_\alpha}} - \varphi\left(\frac{\sigma_\alpha}{\varepsilon}\right)\frac{\partial \lambda^s_{k_\alpha}}{\partial w^+_{k_\alpha}} - \left(1 - \varphi\left(\frac{\sigma_\alpha}{\varepsilon}\right)\right)\frac{\partial \lambda^r_{k_\alpha}}{\partial w^+_{k_\alpha}}\right| \\ &\qquad \leq \frac{1}{\varepsilon}\left|\varphi'\left(\frac{\sigma_\alpha}{\varepsilon}\right)\right| \left|\frac{\partial \sigma_\alpha}{\partial w^+_{k_\alpha}}\right| |\lambda^s_{k_\alpha} - \lambda^r_{k_\alpha}| \leq C|\sigma_\alpha|. \end{aligned} \tag{11.11}$$

We remark that, by (3.15),

$$\left|\frac{\partial u^-}{\partial w^+_{k_\alpha}}\right| = \left|\sum_{i<k_\alpha} r_i^- \frac{\partial W^i}{\partial w^+_{k_\alpha}}\right| \leq C|\sigma_\alpha|^2. \tag{11.12}$$

From (2.5), recalling $\sigma_\alpha < 0$, we get

$$
(11.13)\quad
\begin{aligned}
\frac{\partial \lambda^r_{k_\alpha}}{\partial w^+_{k_\alpha}} &= \frac{1}{|\sigma_\alpha|^2}\frac{\partial \sigma_\alpha}{\partial w^+_{k_\alpha}} \int_{\sigma_\alpha}^0 \lambda_{k_\alpha}(R_h(s)(u^-))\,ds - \frac{1}{|\sigma_\alpha|}\frac{\partial \sigma_\alpha}{\partial w^+_{k_\alpha}}\lambda_{k_\alpha}(u^+) \\
&\quad + \frac{1}{|\sigma_\alpha|}\int_{\sigma_\alpha}^0 D_{u^-}[\lambda_{k_\alpha}(R_h(s)(u^-))]\frac{\partial u^-}{\partial w^+_{k_\alpha}}\,ds \\
&= \frac{1}{|\sigma_\alpha|}\frac{\partial \sigma_\alpha}{\partial w^+_{k_\alpha}}(\lambda^r_{k_\alpha} - \lambda_{k_\alpha}(u^+)) \\
&\quad + \frac{1}{|\sigma_\alpha|}\int_{\sigma_\alpha}^0 D_{u^-}[\lambda_{k_\alpha}(R_h(s)(u^-))]\frac{\partial u^-}{\partial w^+_{k_\alpha}}\,ds.
\end{aligned}
$$

From (11.12), the last term in (11.13) can be estimated by $C|\sigma_\alpha|^2$, hence from (3.17) we get

$$
\left| \frac{\partial \lambda^r_{k_\alpha}}{\partial w^+_{k_\alpha}} - \frac{\lambda^r_{k_\alpha} - \lambda_{k_\alpha}(u^+)}{|\sigma_\alpha|} \right| \le C|\sigma_\alpha|.
$$

From (11.10) now it easily follows that

$$
\left| \frac{\partial \dot{y}_\alpha}{\partial w^+_{k_\alpha}} - \frac{\dot{y}_\alpha - \lambda_{k_\alpha}(u^+)}{|\sigma_\alpha|} \right| \le C|\sigma_\alpha|.
$$

In order to obtain (3.20), it suffices to observe that, since $\alpha \in \mathcal{S}$,

$$
\left| \lambda_{k_\alpha}(u^+) - \lambda^{(h)}_{k_\alpha}(y_\alpha +) \right| \le C\varepsilon^2 \le C|\sigma_\alpha|^2.
$$

In the case $\alpha \in \mathcal{S}'$, recalling (3.21), the estimates in (3.22) are obtained differentiating (3.9) w.r.t. $w^\pm_{k_\alpha}$.

Estimates (3.29), (3.30) Recalling that $\sigma_\alpha = E_{k_\alpha}(u^-, u^+)$, the time derivative of σ_α is computed by

$$
\dot{\sigma}_\alpha = DE_{k_\alpha}(u^-, u^+)\left(-\sum_{i=1}^n w^-_i r^-_i, \sum_{i=1}^n w^+_i r^+_i \right),
$$

where $w_i^- \doteq (\lambda_i^{(h)}(y_\alpha-) - \dot{y}_\alpha)u_x^{i-}$, $w_i^+ \doteq (\dot{y}_\alpha - \lambda_i^{(h)}(y_\alpha+))u_x^{i+}$. We now split the r.h.s. of the above equation in the following way:

$$
\begin{aligned}
\dot{\sigma}_\alpha =& [DE_{k_\alpha}(u^-, u^+) - DE_{k_\alpha}(u^+, u^+)](w_{k_\alpha}^-(r_{k_\alpha}^+ - r_{k_\alpha}^-), 0) \\
&+ DE_{k_\alpha}(u^+, u^+)(w_{k_\alpha}^-(r_{k_\alpha}^+ - r_{k_\alpha}^-), 0) \\
&+ [DE_{k_\alpha}(u^-, u^+) - DE_{k_\alpha}(u^+, u^+)](-w_{k_\alpha}^- r_{k_\alpha}^+, w_{k_\alpha}^+ r_{k_\alpha}^+) \\
&+ DE_{k_\alpha}(u^+, u^+)(-w_{k_\alpha}^- r_{k_\alpha}^+, w_{k_\alpha}^+ r_{k_\alpha}^+) \\
&+ [DE_{k_\alpha}(u^-, u^+) - DE_{k_\alpha}(u^+, u^+)] \left(-\sum_{i \neq k_\alpha} w_i^- r_i^-, \sum_{i \neq k_\alpha} w_i^+ r_i^+ \right) \\
&+ DE_{k_\alpha}(u^+, u^+) \left(-\sum_{i \neq k_\alpha} w_i^- r_i^+, \sum_{i \neq k_\alpha} w_i^+ r_i^+ \right) \\
&+ DE_{k_\alpha}(u^+, u^+) \left(\sum_{i \neq k_\alpha} w_i^-(r_i^+ - r_i^-), 0 \right) \\
\doteq& I_1 + \cdots + I_7.
\end{aligned}
$$

From the definition of $w_i^\pm$ we have that

$$
|w_i^\pm| \leq C|u_x^{i\pm}|, \quad \forall i^\pm, \qquad |w_{k_\alpha}^\pm| \leq C|\sigma_\alpha|\,|u_x^{k_\alpha \pm}|. \tag{11.14}
$$

Recalling that $r_{k_\alpha}^- = r_{k_\alpha}^+ - \sigma_\alpha(\nabla r_{k_\alpha} \cdot r_{k_\alpha})(u^+) + O(\sigma_\alpha^2)$ and $\langle r_{k_\alpha}, \nabla r_{k_\alpha} \cdot r_{k_\alpha} \rangle = 0$, we get

$$
\begin{gathered}
|I_1| \leq C|\sigma_\alpha|^2\,|w_{k_\alpha}^-| \leq C|\sigma_\alpha|^3\,|u_x^{k_\alpha -}|, \\
|I_3| \leq C|\sigma_\alpha|^2(|w_{k_\alpha}^+| + |w_{k_\alpha}^-|) \leq C|\sigma_\alpha|^3(|u_x^{k_\alpha +}| + |u_x^{k_\alpha -}|).
\end{gathered}
$$

Moreover, by (11.3), it easily follows that

$$
|I_2| \leq C|\sigma_\alpha|^2\,|w_{k_\alpha}^-| \leq C|\sigma_\alpha|^3\,|u_x^{k_\alpha -}|.
$$

The term I_4 gives

$$
I_4 = w_{k_\alpha}^- + w_{k_\alpha}^+ = (\lambda_{k_\alpha}^{(h)}(y_\alpha-) - \dot{y}_\alpha)u_x^{k_\alpha -} + (\dot{y}_\alpha - \lambda_{k_\alpha}^{(h)}(y_\alpha+))u_x^{k_\alpha +}.
$$

Concerning I_5, we have that

$$
|I_5| \leq C|\sigma_\alpha| \sum_{i \neq k_\alpha} (|w_i^+| + |w_i^-|). \tag{11.15}
$$

Let us compute $\sum_{i \neq k_\alpha} |w_i^+|$. If $i^\pm \in \mathcal{O}$, by the linearity of the maps W^j we have that

$$
w_i^\pm = \sum_{j^\pm \in \mathcal{I}} \frac{\partial W^i}{\partial w_j^\pm} w_j^\pm, \quad i^\pm \in \mathcal{O},
$$

so that, recalling (3.13)-(3.15) and (11.14),

$$\begin{aligned}\sum_{i\neq k_\alpha}|w_i^+| &\leq \sum_{i<k_\alpha}|w_i^+| + \sum_{i>k_\alpha}\left(\sum_{j^\pm\in\mathcal{I}}\left|\frac{\partial W^i}{\partial w_j^\pm}\right||w_j^\pm|\right)\leq\\ &\leq \sum_{i<k_\alpha}|w_i^+| + \sum_{i>k_\alpha}|w_i^-|(1+C|\sigma_\alpha|) + C|\sigma_\alpha|(|w_{k_\alpha}^+|+|w_{k_\alpha}^-|)\leq\\ &\leq C\sum_{\substack{i^\pm\in\mathcal{I}\\ i\neq k_\alpha}}|u_x^{i\pm}| + C|\sigma_\alpha|^2(|u_x^{k_\alpha+}|+|u_x^{k_\alpha-}|).\end{aligned}$$

Since $\sum_{i\neq k_\alpha}|w_i^-|$ can be estimated in the same way, from (11.15) we obtain

$$|I_5|\leq C|\sigma_\alpha|\sum_{\substack{i^\pm\in\mathcal{I}\\ i\neq k_\alpha}}|u_x^{i\pm}| + C|\sigma_\alpha|^3(|u_x^{k_\alpha+}|+|u_x^{k_\alpha-}|). \tag{11.16}$$

It is easily seen that I_7 satisfies the same estimate, while (11.3) implies that $I_6 = 0$. Now (3.29) follows easily.

In order to obtain the estimate (3.30), it is enough to remark that, if $\alpha\in\mathcal{S}'$, then from (11.7)-(11.8) one has $I_1 + I_2 + I_3 = 0$, and from (3.21) the estimate (11.16) is replaced by the sharper one

$$|I_5|\leq C|\sigma_\alpha|\sum_{\substack{i^\pm\in\mathcal{I}\\ i\neq k_\alpha}}|u_x^{i\pm}|.$$

Formula (3.40) Let $\lambda_h^\pm \doteq \lambda_h^{(h)}(y_\alpha^*(\tau)\pm)$. Fix $\Delta x > 0$, and let $\Delta t \doteq \Delta x/\lambda_h^+$. Let $\theta\mapsto u^\theta$ be a curve which generates the tangent vector v, and assume $u^0 = u$. Assume that $y_\alpha^*(\tau) = 0$. We have that, for $\theta > 0$ small enough,

$$u_h^\theta(\tau+\Delta t,\Delta x) = u_h^\theta\left(\tau, \frac{\lambda_h^+-\lambda_h^-}{\lambda_h^+-\dot y_\alpha}\xi_\alpha\theta\right),\qquad u_h(\tau+\Delta t,\Delta x) = u_h(\tau,0),$$

so that

$$\begin{aligned}&\frac{u_h^\theta(\tau+\Delta t,\Delta x)-u_h(\tau+\Delta t,\Delta x)}{\theta}=\\ &\qquad=\frac{u_h^\theta\left(\tau,\frac{\lambda_h^+-\lambda_h^-}{\lambda_h^+-\dot y_\alpha}\xi_\alpha\theta\right)-u_h^\theta(\tau,0)}{\theta}+\frac{u_h^\theta(\tau,0-)-u_h(\tau,0-)}{\theta}\\ &\qquad\approx u_x^h(\tau,0-)\cdot\frac{\lambda_h^+-\lambda_h^-}{\lambda_h^+-\dot y_\alpha}\xi_\alpha + v_h(\tau,0-).\end{aligned}$$

Passing to the limit for $\theta\to 0$ and then for $\Delta x\to 0$ we thus obtain

$$v_h^+ = v_h^- + \xi_\alpha u_x^{h-}\cdot\frac{\lambda_h^+-\lambda_h^-}{\lambda_h^+-\dot y_\alpha}.$$

Recalling (3.28), we have that

$$u_x^{h-}\cdot\frac{\lambda_h^+-\lambda_h^-}{\lambda_h^+-\dot y_\alpha} = u_x^{h-}\left(1-\frac{\lambda_h^--\dot y_\alpha}{\lambda_h^+-\dot y_\alpha}\right) = u_x^{h-}-u_x^{h+},$$

hence the first relation in (3.40) follows. The simpler relation $v_i^+ = v_i^-$ for every $i\neq h$ can be proved in the same way.

Estimate (3.58) We recall that at a point of shock y_α, $\alpha\in\mathcal{S}\cup\mathcal{S}'$, the equations

(3.12) are satisfied with $w_i^\pm$ defined as in (3.39). Now let $i \neq k_\alpha$. Assume, for example, $i > k_\alpha$, so that $i^+ \in \mathcal{O}$. From (3.12) we have that

$$\xi_\alpha u_x^{i+} + v_i^+ = \sum_{\substack{j^\pm \in \mathcal{I} \\ j \neq i}} \frac{\partial W^i}{\partial w_j^\pm}(\xi_\alpha u_x^{j\pm} + v_j^\pm) + \left(\frac{\partial W^i}{\partial w_i^-} - 1\right)(\xi_\alpha u_x^{i-} + v_i^-) + \xi_\alpha u_x^{i-} + v_i^- .$$

From (3.13)-(3.15) we get

$$\begin{aligned} |v_i^+ - v_i^-| \leq & |\xi_\alpha|\,|u_x^{i+} - u_x^{i-}| + C|\sigma_\alpha| \sum_{j^\pm \in \mathcal{I}} |\xi_\alpha u_x^{j\pm} + v_j^\pm| \\ & + C|\sigma_\alpha|^2 \left(|\xi_\alpha u_x^{k_\alpha +} + v_{k_\alpha}^+| + |\xi_\alpha u_x^{k_\alpha -} + v_{k_\alpha}^-|\right). \end{aligned}$$

The same estimate holds if $i < k_\alpha$. Recalling the definition (3.35) and (3.54) of Λ_α and Ψ_α we get (3.58).

Estimate (3.59) We have that

$$\begin{aligned} &\left|(\lambda_{k_\alpha}^{(h)}(y_\alpha-) - \dot{y}_\alpha)\eta^- + (\dot{y}_\alpha - \lambda_{k_\alpha}^{(h)}(y_\alpha+))\eta^+ - |\sigma_\alpha| D\lambda_{k_\alpha}^{(h)}(u^-, u^+)(\eta^- r_{k_\alpha}^-, \eta^+ r_{k_\alpha}^+)\right| \\ &\qquad \leq |\sigma_\alpha|(|\eta^-| I_1 + |\eta^+| I_2), \end{aligned}$$

where

$$I_1 \doteq \left| \frac{\lambda_{k_\alpha}^{(h)}(y_\alpha-) - \dot{y}_\alpha}{|\sigma_\alpha|} - D\lambda_{k_\alpha}^{(h)}(u^-, u^+)(r_{k_\alpha}^-, 0)\right|,$$

$$I_2 \doteq \left| \frac{\dot{y}_\alpha - \lambda_{k_\alpha}^{(h)}(y_\alpha-)}{|\sigma_\alpha|} - D\lambda_{k_\alpha}^{(h)}(u^-, u^+)(0, r_{k_\alpha}^+)\right|.$$

Observe that, from

$$\left|\dot{y}_\alpha - \lambda_{k_\alpha}^{(h)}(u^-, u^+)\right| \leq C|\sigma_\alpha|^2,$$

one obtains

$$\left| \frac{\partial \dot{y}_\alpha}{\partial w_{k_\alpha}^+} - D\lambda_{k_\alpha}^{(h)}(u^-, u^+)\left(\frac{\partial u^-}{\partial w_{k_\alpha}^+}, \frac{\partial u^+}{\partial w_{k_\alpha}^+}\right)\right| \leq C|\sigma_\alpha|. \tag{11.17}$$

Now one has

$$\begin{aligned} D\lambda_{k_\alpha}^{(h)}(u^-, u^+)&\left(\frac{\partial u^-}{\partial w_{k_\alpha}^-}, \frac{\partial u^+}{\partial w_{k_\alpha}^-}\right) = \\ &= D\lambda_{k_\alpha}^{(h)}(u^-, u^+)\left(r_{k_\alpha}^- + \sum_{j<k_\alpha} r_j^- \frac{\partial W^j}{\partial w_{k_\alpha}^-}, \sum_{j>k_\alpha} r_j^+ \frac{\partial W^j}{\partial w_{k_\alpha}^-}\right), \end{aligned}$$

so that, from (3.15),

$$\begin{aligned} &\left| D\lambda_{k_\alpha}^{(h)}(u^-, u^+)\left(\frac{\partial u^-}{\partial w_{k_\alpha}^+}, \frac{\partial u^+}{\partial w_{k_\alpha}^+}\right) - D\lambda_{k_\alpha}^{(h)}(u^-, u^+)(r_{k_\alpha}^-, 0)\right| \leq \\ &\qquad \leq C \sum_{j^\pm \in \mathcal{O}} \left|\frac{\partial W^j}{\partial w_{k_\alpha}^-}\right| \leq C|\sigma_\alpha|^2. \end{aligned} \tag{11.18}$$

From (11.17) and (11.18) one obtains

$$\left| \frac{\partial \dot{y}_\alpha}{\partial w^-_{k_\alpha}} - D\lambda^{(h)}_{k_\alpha}(u^-,u^+)(r^-_{k_\alpha},0) \right| \leq C|\sigma_\alpha|,$$

so that, recalling (3.19), it follows that

$$I_1 \leq \left| \frac{\lambda^{(h)}_{k_\alpha}(y_\alpha -) - \dot{y}_\alpha}{|\sigma_\alpha|} - \frac{\partial \dot{y}_\alpha}{\partial w^-_{k_\alpha}} \right| + \left| \frac{\partial \dot{y}_\alpha}{\partial w^-_{k_\alpha}} - D\lambda^{(h)}_{k_\alpha}(u^-,u^+)(r^-_{k_\alpha},0) \right| \leq C|\sigma_\alpha|,$$

and a similar estimate holds for I_2.

In the case $\alpha \in S'$, from the definition (3.9) it easily follows that $I_1 = I_2 = 0$.

Estimates (3.66), (3.67) The first estimate in (3.66) is a consequence of the fact

$$|\lambda^{(h)}_h(y^*_\alpha +) - \lambda^{(h)}_h(y^*_\alpha -)| \leq C \cdot [\text{strength of external waves}] \leq C\eta.$$

Concerning the second estimate, by (3.28) we have that either $u^{h-}_x = u^{h+}_x = 0$ or $u^{h-}_x \cdot u^{h+}_x \neq 0$. In the first case the estimate is trivial. Assume now $0 < |u^{h-}_x| \leq |u^{h+}_x|$. Recalling (3.28), by strict hyperbolicity we get

$$\begin{aligned} |v_h(y_\alpha +) - v_h(y_\alpha -)| =& |\xi_\alpha|\, |u^{h+}_x - u^{h-}_x| \\ =& |\xi_\alpha|\, |u^{h-}_x| \frac{|\lambda^{(h)}_h(y^*_\alpha +) - \lambda^{(h)}_h(y^*_\alpha -)|}{|\lambda^{(h)}_h(y^*_\alpha +) - \dot{y}_\alpha|} \leq C|\xi_\alpha|\, |u^{h-}_x|\eta. \end{aligned}$$

The estimates (3.67) follows from the strict hyperbolicity and $|\sigma_\alpha| \geq \varepsilon$.

Estimate (3.72) The first relation is an easy consequence of the coincidence of shock and rarefaction curves.

Concerning the second relation, from (5.10) in [**B4**] we have that

(11.19)
$$\xi(\tau+) = \frac{\xi'(\tau-)(\lambda^{(h)}_h(\omega_{h-1},\omega_h) - \dot{y}'(\tau-)) - \xi''(\tau-)(\lambda^{(h)}_h(\omega_{h-1},\omega_h) - \dot{y}'(\tau-))}{\dot{y}'(\tau-) - \dot{y}''(\tau-)}.$$

Since shock and rarefaction curves coincide, one has

$$\lambda^{(h)}_h(\omega_{h-1},\omega_h) = \frac{\sigma'(\tau-)\dot{y}'(\tau-) + \sigma''(\tau-)\dot{y}''(\tau-)}{\sigma'(\tau-) + \sigma''(\tau-)}.$$

Substituting in (11.19) we now obtain the second relation in (3.72).

Bibliography

[B-B] P. Baiti and A. Bressan, Lower semicontinuity of weighted path length in BV, *Geometrical Optics and Related Topics*, F. Colombini and N. Lerner Eds., Birkhäuser, 1997, 31-58.

[B-J] P. Baiti and H. K. Jenssen, On the front tracking algorithm, *J. Math. Anal. Appl.* **217** (1998), 395-404.

[B1] A. Bressan, Contractive metrics for nonlinear hyperbolic systems, *Indiana Univ. Math. J.* **37** (1988), 409-421.

[B2] A. Bressan, Global solutions to systems of conservation laws by wave-front tracking, *J. Math. Anal. Appl.* **170** (1992), 414-432.

[B3] A. Bressan, A contractive metric for systems of conservation laws with coinciding shock and rarefaction curves. *J. Differential Equations* **106** (1993), 332-366.

[B4] A. Bressan, A locally contractive metric for systems of conservation laws, *Ann. Scuola Norm. Sup. Pisa* **IV-22** (1995), 109-135.

[B5] A. Bressan, The unique limit of the Glimm scheme, *Arch. Rational Mech. Anal.* **130** (1995), 205-230.

[B6] A. Bressan, *Lecture Notes on Systems of Conservation Laws*, S.I.S.S.A., Trieste 1996.

[B-C1] A. Bressan and R. M. Colombo, The semigroup generated by 2×2 conservation laws, *Arch. Rational Mech. Anal.* **133** (1995), 1-75.

[B-C2] A. Bressan and R. M. Colombo, Unique solutions of 2×2 conservation laws with large data, *Indiana Univ. Math. J.* **44** (1995), 677-725.

[B-C3] A. Bressan and R. M. Colombo, Decay of positive waves in nonlinear systems of conservation laws, *Ann. Scuola Norm. Sup. Pisa Cl. Sci. (4)* **26** (1998), 219-245.

[B-G] A. Bressan and P. Goatin, Oleinik type estimates and uniqueness for $n \times n$ conservation laws, to appear on *J. Differential Equat.*

[B-LF1] A. Bressan and P. LeFloch, Uniqueness of weak solutions to systems of conservation laws, *Arch. Rational Mech. Anal.* **140** (1997), 301-317.

[B-LF2] A. Bressan and P. LeFloch, Structural stability of solutions to systems of conservation laws, to appear on *Indiana Univ. J. Math.*

[B-M1] A. Bressan and A. Marson, A variational calculus for shock solutions of systems of conservation laws, *Comm. Part. Diff. Equat.* **20** (1995), 1491-1552.

[B-M2] A. Bressan and A. Marson, Error bounds for a deterministic version of the Glimm scheme, *Arch. Rational Mech. Anal.* **142** (1998), 155-176.

[Cl] F. H. Clarke, *Optimization and Nonsmooth Analysis*, Wiley, New York, 1983.

[C] M. Crandall, the semigroup approach to first-order quasilinear equations in several space variables, *Israel J. Math.* **12** (1972), 108-132.

[D] C. Dafermos, Polygonal approximations of solutions of the initial value problem for a conservation laws, *J. Math. Anal. Appl.* **38** (1972), 33-41.

[DP1] R. DiPerna, Singularities of solutions of nonlinear hyperbolic systems of conservation laws, *Arch. Rational Mech. Anal.* **60** (1975), 75-100.

[DP2] R. DiPerna, Global existence of solutions to nonlinear hyperbolic systems of equations, *J. Differential Equations* **20** (1976), 187-212.

[E-G] L. C. Evans and R. F. Gariepy, *Measure Theory and Fine properties of Functions*, CRC Press, Boca Raton, 1992.

[G] J. Glimm, Solutions in the large for nonlinear hyperbolic systems of equations, *Comm. Pure Appl. Math.* **18** (1965), 697-715.

[K] S. Kruzkov, First order quasilinear equations with several space variables, *Math. USSR Sbornik* **10** (1970), 217-243.

[Lx] P. Lax, Hyperbolic systems of conservation laws II, *Comm. Pure Appl. Math.* **10** (1957), 537-566.

[L] T. P. Liu, The deterministic version of the Glimm scheme, *Comm. Math. Phys.* **57** (1977), 135-148.

[O] O. Oleinik, Discontinuous solutions of nonlinear differential equations, *Usp. Mat. Nauk.* **12** (1957), 3-73; English transl. in *Amer. Math. Soc. Transl. Ser. 2,* **26**, 95-172.

[R] N. H. Risebro, A front-tracking alternative to the random choice method, *Proc. Amer. Math. Soc.* **117** (1993), 1125-1139.

[Sch] M. Schatzman, Continuous Glimm functionals and uniqueness of solutions of the Riemann problem, *Indiana Univ. Math. J.* **34** (1985), 533-589.

[Sm] J. Smoller, *Shock Waves and Reaction-Diffusion Equations*, Springer-Verlag, New York, 1983.

[T1] B. Temple, Systems of conservation laws with invariant submanifolds, *Trans. Amer. Math. Soc.* **280** (1983), 781-795.

[T2] B. Temple, No L^1-contractive metrics for systems of conservation laws, *Trans. Amer. Math. Soc.* **288** (1985), 471-480.

Editorial Information

To be published in the *Memoirs*, a paper must be correct, new, nontrivial, and significant. Further, it must be well written and of interest to a substantial number of mathematicians. Piecemeal results, such as an inconclusive step toward an unproved major theorem or a minor variation on a known result, are in general not acceptable for publication. *Transactions* Editors shall solicit and encourage publication of worthy papers. Papers appearing in *Memoirs* are generally longer than those appearing in *Transactions* with which it shares an editorial committee.

As of March 31, 2000, the backlog for this journal was approximately 7 volumes. This estimate is the result of dividing the number of manuscripts for this journal in the Providence office that have not yet gone to the printer on the above date by the average number of monographs per volume over the previous twelve months, reduced by the number of issues published in four months (the time necessary for preparing an issue for the printer). (There are 6 volumes per year, each containing at least 4 numbers.)

A Copyright Transfer Agreement is required before a paper will be published in this journal. By submitting a paper to this journal, authors certify that the manuscript has not been submitted to nor is it under consideration for publication by another journal, conference proceedings, or similar publication.

Information for Authors and Editors

Memoirs are printed by photo-offset from camera copy fully prepared by the author. This means that the finished book will look exactly like the copy submitted.

The paper must contain a *descriptive title* and an *abstract* that summarizes the article in language suitable for workers in the general field (algebra, analysis, etc.). The *descriptive title* should be short, but informative; useless or vague phrases such as "some remarks about" or "concerning" should be avoided. The *abstract* should be at least one complete sentence, and at most 300 words. Included with the footnotes to the paper, there should be the 2000 *Mathematics Subject Classification* representing the primary and secondary subjects of the article. This may be followed by a list of *key words and phrases* describing the subject matter of the article and taken from it. A list of the numbers may be found in the annual index of *Mathematical Reviews*, published with the December issue starting in 1990, as well as from the electronic service e-MATH [**telnet e-MATH.ams.org** (or **telnet 130.44.1.100**). Login and password are **e-math**]. For journal abbreviations used in bibliographies, see the list of serials on the web at `http://www.ams.org/msnhtml/serials-list/annser_frames.html`. When the manuscript is submitted, authors should supply the editor with electronic addresses if available. These will be printed after the postal address at the end of each article.

Electronically prepared papers. The AMS encourages submission of electronically prepared papers in AMS-TeX or AMS-LaTeX. The Society has prepared author packages for each AMS publication. Author packages include instructions for preparing electronic papers, the *AMS Author Handbook*, samples, and a style file that generates the particular design specifications of that publication series for both AMS-TeX and AMS-LaTeX.

Authors with FTP access may retrieve an author package from the Society's Internet node `e-MATH.ams.org` (130.44.1.100). For those without FTP access, the author package can be obtained free of charge by sending e-mail to `pub@ams.org` (Internet) or from the Publication Division, American Mathematical Society, P.O. Box 6248, Providence, RI 02940-6248. When requesting an author package, please specify $\mathcal{A}\mathcal{M}\mathcal{S}$-TEX or $\mathcal{A}\mathcal{M}\mathcal{S}$-LATEX, Macintosh or IBM (3.5) format, and the publication in which your paper will appear. Please be sure to include your complete mailing address.

Submission of electronic files. At the time of submission, the source file(s) should be sent to the Providence office (this includes any TEX source file, any graphics files, and the DVI or PostScript file).

Before sending the source file, be sure you have proofread your paper carefully. The files you send must be the EXACT files used to generate the proof copy that was accepted for publication. For all publications, authors are required to send a printed copy of their paper, which exactly matches the copy approved for publication, along with any graphics that will appear in the paper.

TEX files may be submitted by email, FTP, or on diskette. The DVI file(s) and PostScript files should be submitted only by FTP or on diskette unless they are encoded properly to submit through e-mail. (DVI files are binary and PostScript files tend to be very large.)

Files sent by electronic mail should be addressed to the Internet address `pub-submit@ams.org`. The subject line of the message should include the publication code to identify it as a Memoir. TEX source files, DVI files, and PostScript files can be transferred over the Internet by FTP to the Internet node `e-math.ams.org` (130.44.1.100).

Electronic graphics. Figures may be submitted to the AMS in an electronic format. The AMS recommends that graphics created electronically be saved in Encapsulated PostScript (EPS) format. This includes graphics originated via a graphics application as well as scanned photographs or other computer-generated images.

If the graphics package used does not support EPS output, the graphics file should be saved in one of the standard graphics formats—such as TIFF, PICT, GIF, etc.—rather than in an application-dependent format. Graphics files submitted in an application-dependent format are not likely to be used. No matter what method was used to produce the graphic, it is necessary to provide a paper copy to the AMS.

Authors using graphics packages for the creation of electronic art should also avoid the use of any lines thinner than 0.5 points in width. Many graphics packages allow the user to specify a "hairline" for a very thin line. Hairlines often look acceptable when proofed on a typical laser printer. However, when produced on a high-resolution laser imagesetter, hairlines become nearly invisible and will be lost entirely in the final printing process.

Screens should be set to values between 15% and 85%. Screens which fall outside of this range are too light or too dark to print correctly.

Any inquiries concerning a paper that has been accepted for publication should be sent directly to the Editorial Department, American Mathematical Society, P. O. Box 6248, Providence, RI 02940-6248.

Editors

Selected Titles in This Series

(Continued from the front of this publication)